Theory and Approaches of Unascertained Group Decision-Making

Systems Evaluation, Prediction, and Decision-Making Series

RECENTLY PUBLISHED

Theory and Approaches of Unascertained Group Decision-Making

Jianjun Zhu

CRC Press
Taylor & Francis Group
Boca Raton   London   New York

CRC Press is an imprint of the
Taylor & Francis Group, an **informa** business
AN AUERBACH BOOK

CRC Press
Taylor & Francis Group
6000 Broken Sound Parkway NW, Suite 300
Boca Raton, FL 33487-2742

First issued in paperback 2016

ISBN 13: 978-1-138-19898-2 (pbk)
ISBN 13: 978-1-4200-8750-5 (hbk)

Library of Congress Cataloging-in-Publication Data

Zhu, Jianjun.
    Theory and approaches of unascertained group decision-making / Jianjun Zhu.
        p. cm. -- (Systems evaluation, prediction, and decision-making series)
    Includes bibliographical references and index.
    ISBN 978-1-4200-8750-5 (hardback)
    1. Decision making. 2. System theory. I. Title.

T57.95.Z48 2012
302.3--dc23   2012005276

Visit the Taylor & Francis Web site at
http://www.taylorandfrancis.com

and the CRC Press Web site at
http://www.crcpress.com

Contents

PREFACE   vii

ABOUT THE AUTHOR   ix

ACKNOWLEDGMENTS   xi

ABSTRACT   xiii

CHAPTER 1   INTRODUCTION   1

CHAPTER 2   DECISION-MAKING METHOD OF INTERVAL NUMBER RECIPROCAL AND COMPLEMENTARY COMPARISON MATRIX   13

CHAPTER 3   DECISION-MAKING METHOD OF UNASCERTAINED NUMBER COMPARISON MATRIX   49

CHAPTER 4   DECISION-MAKING METHOD OF THREE-POINT INTERVAL NUMBER COMPARISON MATRIX   65

CHAPTER 5   DECISION-MAKING METHOD OF THE LINGUISTIC COMPARISON MATRIX   91

CHAPTER 6   AGGREGATION METHOD ON MULTIPLE STYLE PREFERENCE FOR GROUP DECISION MAKING   125

CHAPTER 7   AGGREGATION METHOD OF MULTIPLE STAGES FOR GROUP DECISION MAKING   177

v

CHAPTER 8   AN EXTENSION OF TOPSIS WITH MULTIPLE-
             STAGES FUZZY LINGUISTIC EVALUATION FOR
             GROUP DECISION MAKING   201

CONCLUSION AND SUMMARY   227
REFERENCES   231
INDEX   239

Preface

The famous economist H. A. Simon, one of the founders of scientific management, said that management is decision making. Decision making is the evaluation and selection process by the decision maker (DM) that aims to achieve some purpose or to accomplish a task. With the development of society and the great increase of knowledge and information, more decision-making problems involve a number of decision makers. The subjective preference of DMs reflects a particular analysis, thinking process, and cognitive activity of the decision-making problem. Because of the uncertainty of the decision-making environment, DMs tend to express their preference with interval numbers, fuzzy numbers, and linguistic variables. As a result, several uncertain preference styles, such as judgment matrix, utility value, and preference ordering value of interval numbers, fuzzy numbers and linguistic term sets are given by DMs. Owing to many of assessment factors involved in complex decision-making problems, the difference of preferences, and the impact of the internal and external environment, it is often difficult to aggregate information in the group decision-making process. As a result, this book studies mainly group aggregation methods based on uncertainty preference information.

The book is constructed as follows. The studies on group decision making are reviewed in Chapter 1. The consistency measuring and ranking methods of interval number reciprocal judgment matrix and interval number complementary judgment matrix are discussed in Chapter 2. An unascertained number preference and a three-point interval number preference are presented in Chapters 3 and 4, and their consistencies are defined and the ranking methods of the alternative are developed. The linguistic preference is studied in Chapter 5, and two consistency definitions have been put forward. The aggregating methods of several uncertain preferences are discussed in Chapter 6. The multistage aggregating model of uncertain preference is studied in Chapter 7. An aggregating model of multistage linguistic information based on TOPSIS is proposed in Chapter 8.

The research referred to in this book was supported by the National Natural Science Foundation of China; the Social Science Foundation of China, Jiangsu Province; the Soft Science Foundation of China, Jiangsu Province; the Postdoctoral Programs Foundation of China, Jiangsu Province; the Doctoral Programs Foundation of Chinese National Educational Ministry; the Social Science Foundation of Chinese National Educational Ministry; the Research Foundation for the Excellent and Creative Teamwork in Science and Technology in Higher Institutions and Distinguished Professors of Jiangsu Province; and the Foundation for the Master's and Doctoral Programs in the Eleventh Five-Year Plan in Nanjing University of Aeronautics and Astronautics. In the writing of this book, the author consulted widely, referred to research by many scholars, and was helped greatly by Professor Jeffrey Forrest and Professor Sifeng Liu. I wish to thank them all.

Any errors or omissions that may be pointed out by readers and experts in this field will be appreciated.

About the Author

Jianjun Zhu received a B.A. and an M.A. in coal mining engineering from the Liaoning Technical University, Fuxin, Liaoning, China, and a Ph.D. in system engineering from the Northeastern University, Shenyang, Liaoning, China. He is currently a visiting professor at Waterloo University, Canada.

He is a chair professor with the Department of Management Science and Engineering at the University of Nanjing University of Aeronautics and Astronautics, Nanjing, Jiangsu, China. Previously, he was a Postdoctoral Fellow with the Department of Management Science and Engineering, University of Nanjing University of Aeronautics and Astronautics. He is a director of the Chinese Society for Optimization, Overall Planning and Economic Mathematics (CSOOPEM), and the Grey Systems Society of China. His major research interests are the group preference aggregation, gray systems theory, and uncertain decision-making analysis. The main application areas of these technologies are vendor rank and selection, government policy evaluation, and complex system design and optimization. He is the author of three edited books and close to 40 papers.

Acknowledgments

Relevant research was supported by the National Natural Science Foundation of China (Nos. 70701017, 90924022, 70971064) and the Special Project of Basic Science Research of Nanjing University of Aeronautics and Astronautics (No. NS2010209). Additionally, the author would like to acknowledge the partial support of the Science Foundation for the Excellent and Creative Group in Science and Technology of Nanjing University of Aeronautics and Astronautics and Jiangsu Province (No. Y0553-091), the Foundation for Key Research Base of Philosophy and Social Science in Colleges and Universities of Jiangsu Province, and the Foundation for Outstanding Teaching Group of China (No. 10td128). In writing this book, the author consulted widely, referred to research of many scholars, and received a lot of help from Professor Jeffrey Forrest and Professor Sifeng Liu. I wish to thank them all.

Abstract

Decision making reflects the thought process of analysis and disposal, which often depends on subjective preference. The frequently preferred styles include comparison matrix, utility value, rank value, linguistic value, etc. The decision maker often adopts interval numbers and fuzzy numbers to express his preference due to environmental uncertainty. In a group decision-making process, individual preferences cannot be accommodated easily since a complex decision may involve numerous evaluation factors and decision makers. Various views among decision makers may lead to unrealistic results and ineffective actions. Moreover, information system technology developments require practical, adaptive, and intelligent group decision-making systems and methods to aggregate the various preferences must be developed.

This book focuses on three aspects of decision making. First, a consistency definition for uncertain preferences is developed and a modification method for handling inconsistent preferences is suggested.

A mechanism for dealing with uncertainty preferences in group decision making is discussed. In addition, a new uncertainty preference style is introduced—an uncertainty comparison matrix consisting of three parameters is adopted to express the preference of the decision maker.

Second, the aggregation of the decision makers' multiple uncertainty preferences is studied. The satisfaction indicators of group decision making in the background of uncertainty are constructed to predict and assess the effect of group decision making. For the uncertain multistructure preference information's inherent characteristics, the judgment differences are studied based on the satisfaction degree of decision making and the indicators of consistency. The ranking methods for uncertain multi-style preferences are studied in order to achieve group preference based on a maximally consistent view.

Third, the aggregating method of the timing characteristics' multistructure uncertain preference information are studied based on the process of the timing characteristics' decision making. The differences in decision makers' preferences in various stages are analyzed. The algorithms for finding the inconsistent judgments of each stage are suggested. Methods for determining the weight of the process of decision making and the decision group are studied based on the satisfaction degree indicators of decision-makers' uncertain preferences in various stages. The aggregating method of the timing characteristics' multistructure uncertain preference information in various stages is proposed.

Keywords: group decision making; aggregation; uncertainty; preference.

1

INTRODUCTION

1.1 Background and Purpose

The key processes of decision making include the evaluation and selection of people. Related theoretical methods are based on the subject's cognitive activities, which reflect a speculative analysis and treatment process. Therefore, decision making often depends on decision makers' subjective preferences; common preference structures are judgment matrix, utility value, preference ordering value, linguistic value, and so on. Because of the uncertainty of the decision-making environment, decision makers tend to express in the form of preference, such as interval numbers, fuzzy numbers, and linguistic variables. This book studies mainly group decision-making methods and group aggregation methods based on uncertainty preference information. The famous economist H. A. Simon [1], one of the founders of scientific management, said that "management is decision making," and decision making is a conscious and optional action process to achieve some purpose or to accomplish a task. With the development of society, science, and technology, the amount of knowledge and information has greatly increased, wherein more and more decision-making problems involve a number of decision makers. Decision making encompasses the three following aspects of complexity:

1. Uncertain property of the preference structure: uncertainty is absolute, whereas certainty is relative. Decision-making groups always cognize, predict, and judge things in complex and dynamic decision space. Even if the preference is deterministic, there are inevitably some uncertainties. In many cases, the uncertain preference is more relevant for decision making and is usually adopted by decision makers.

2. Multiform property of the preference structure: Extensive application of internet technology makes it possible for many decision makers to be involved in complex decision-making problems. Because of differences in social and cultural backgrounds, life experience, work experience, psychological quality, judging level, the external environment, personal preferences, and so on, different forms of preference information may be given on the same decision by different decision makers to solve the same problem, even in the same space-time. At the same time, due to the incompleteness and asymmetry of decision-making information, and the complexity of the decision-making object structure, decision makers often supply multiple uncertain preferences.

3. Properties of dynamic and multiple stages in the decision-making process: Decision makers' awareness of objective things follows the law of progressive approach wherein conditions are changing and in constant development, so a comprehensive, contacted, dynamic perspective is needed in the decision-making process. In addition, decision makers often need to pay attention to the characteristics of multiple stages in the decision-making process for comprehensive assessment; for example, postevaluation requires a combination of prefeasibility study, feasibility study, design, engineering development, and other stages with which those stages interact, and different forms of preference may be given by decision makers in different stages. A score form involves a numerical evaluation (e.g., a score of 98 points). A voting form requests a qualitative evaluation (e.g., yes or no, qualified or unqualified).

Due to the legion of assessment factors involved in complex decision-making problems, and the difference between decision makers' awareness of things, and the impact of internal and external environment in the decision-making process, it is often difficult to effectively gather group preference information in the group decision-making process, which leads to decentralized decision-making advice and even contrary evaluation findings and seriously affects the decision-making process, thereby adversely affecting the "fighters." In addition, with the development of an intelligent decision support system in the combination of communications and computer technology, the open comprehensive integrated discussion system needs to improve the practical applicability and flexibility of group decision-making technology. Therefore, it is necessary to determine how to effectively aggregate uncertain preference with different structures supplied by decision makers. The background above is the basis of this book.

1.2 Review of the Research

1.2.1 Review of the Uncertain Decision-Making and Group Decision-Making Method

Decision making became the universally accepted expert research in the academic field, which can be traced back to statistical decision theory in the 1950s, and L. J. Savage, Abraham Wald, R. A. Fisher, and other scholars are representative. H. Raiffa and R. O. Schlaifer developed the Bayesian statistical theory, and Harvard Business School researchers represented by H. Raiffa have applied this theory to practical business problems, which contributed to the application of statistical decision theory. In 1966, Howard first proposed the "decision analysis" concept at the 4th International Operations Conference, which has become synonymous with scientific research on decision making.

With increasingly complex human social activities, research on practical problems involves larger and more complex systems that are characterized by more prominent uncertainty, and deterministic description of classical methods becomes powerless. After many years of research, fuzzy math, gray systems, unascertained mathematics, rough sets, and other mathematical theories of uncertainty have been developed, and the application in decision-making also contributed to decision theory being perfected. Probability theory and mathematical statistics can deal with random information. Zhou [1a] and Hu [2] discussed decision-making questions based on random preference. Hahn [3] proposed a random multiple attribute decision-making method based on Bayesian theory. Sabbudin [4] proposed a new algorithm to solve multistage decision-making problem application of Markov. Hryniewicz [5] studied the testing problem of random decision making. L. A. Zadeh [5a, 5b] proposed that fuzzy set–based fuzzy mathematics can express fuzzy information. Gu and Zhu [6] proposed a fuzzy multiattribute decision-making method based on feature vector space. Tang et al. [7] studied the applications of the fuzzy theory in aggregate production planning decision making. Walk and Rutkowski [8] proposed an application model of the fuzzy decision support system. Because fuzzy decision making comes down to comparing fuzzy sets, many papers have been published that proposed ranking methods of fuzzy sets [9–11]. Deng [12] and Liu et al. [13] proposed gray system theory and a mathematical approach to deal with the gray information. Liu [14] reviewed the research progress of the gray system theory. Wang [15] proposed an unascertained information-processing method. Liu et al. [16] summarized the theoretical method and application areas of unascertained mathematics. Other typical findings include: Parkan and Wang [17] and Parkan et al. [18] determined an effective strategy making use of incomplete probability information. Researchers [19, 20, 21] have studied uncertain multiattribute group decision making and information assembly problems making use of evidence theory. Qiu [22] and Chang et al. [23] used information entropy to deal with uncertain decision-making problems. Zhau [24] described uncertainty making use of set pair analysis. Pawlak [25] proposed a rough-set theory to deal with the imprecise problem. Wang et al. [26] proposed a universal gray-set theory. Liu et al. [27] described the application approach of prospect theory. Uncertain decision-making theory grows in its association with uncertainty mathematics. In recent years, some new uncertainty mathematical theories have been introduced to the decision domain, and comprehensive integration application of a variety of uncertainty mathematical theories in complex decision-making environments will become a research hotspot.

The uncertainty and complexity of decision-making problems lead to extensive use of group decision making, and the earliest group decision-making studies date back to medieval social choice theory. K. J. Arrow [28], a Nobel Laureate in Economics, proposed the Arrow impossibility theorem, which led to modern social choice theory and became a classic conclusion of group decision making. In 1975, group decision making was proposed as a clear concept by Bacharach [29] and Keeney and Kirkwood [30], followed by a wide range of academic attention. Since 2002, more than 50 researchers have chosen group decision theory as the direction of their Ph.D. theses and achieved progress in preference theory, group utility theory, social choice theory, negotiation method theory, voting theory, game theory generally, expert evaluation and analysis, quantitative factor assembly, random and fuzzy group decision theory, economic equilibrium theory, group decision support systems, etc. [31–37]. Though research on group decision theory and methods has not formed a unified and rigorous system [38, 39], the next focus of research can be summarized in the following ways:

1. Dynamic process of group decision making: the existing research is still very weak.
2. Organization theory of group decision making: The organization process of group decision making plays an important role in the decision results, and it is an inevitable requirement for scientific group decision making to strengthen research on organization processes. In addition, organizational structure and management are complex in decision-making groups.
3. Communications and impacts among group members: Exchange of information can enhance the judging ability of group members but also easily lead to correlation among group members and the convergence effect. Especially in complex decision environments, uncertain preference may affect correlation and convergence effect among members and complicate decision making.
4. Selecting preference aggregation rules and building an aggregation model: specific decision problems, decision-making groups, and spatial and temporal characteristics often lead to different assembly standards.
5. Strategic behavior in group decision making: Decision makers often represent their own interest groups in various degrees; to a certain degree conflict of interest and interest relevance exist among group organizations, inevitably involving decision-making rivalry and games.
6. The development and application of a group decision support system platform: At present, enterprises and organizations are increasingly demanding for application software of decision making, but a truly effective software is rare [40, 41]. In particular, software platforms that respond to the complicated decision-making environment is in severe shortage.

1.2.2   Review of the Complexity of Group Decision-Making and Related Studies

Many new features presented in group decision making are due to the complexity of the decision-making process. The above-mentioned research related to the complexity of the group decision-making processes. Following Arrow's impossibility theorem, breakthroughs in group decision theory have been rare. As a result, people still recognize the complexity in group decision making insufficiently and lack effective solution strategies. The complexity of the group decision-making process abounds in areas such as multilayer, multifactor, changeable, nonlinear interaction, complexity of time and space evolution, randomness, and other characteristics [42, 43]. The causes can be summarized as follows: things run irregularly, which is completely random; things run regularly, but people still have not found the objective law or have erroneous cognition; things run regularly, but people still cannot accurately grasp and handle the framework of the existing system of theoretical knowledge.

Over the years, research on the complex problem has been a hot issue for scholars who have described the complexity from the entropy, information, fractal dimension, thermodynamic depth, and time and space [43]. The U.S. National Science Foundation published "2006–2011 Strategic Plan," whose key research areas are also involved in the modeling of complex systems and other new cross-disciplines. Early research on Marca Loch and Pitt Heights's neural networks, Neumann's cellular automata, and Wiener's cybernetics made a substantive contribution on the complexity research. In 1980s, the Santa Fe Institute made outstanding contributions to complexity research, and Eisenstein summed up recent progress in complexity research, such as multiscale approaches in the understanding of the complex behavior, nonlinear mechanics and dynamics of the network, agent-based modeling, economic and social complex systems modeling, and so on. Since the 1990s, a comprehensive integration method based on the combination of people's wisdom and high-performance computers, qualitative and quantitative, came into being [44–46]. In the case of complex situations, experts gave judgments and assumptions (and various data and information), and then sublimed the qualitative understandings to quantitative, conclusions through the computer's process; but the modeling method of interactions among main subjects of the system and interactions has still not been effectively addressed, as there is no literature discussing the algorithms of information integration processing over the computer.

1.2.3 Review of Preference Aggregation for Group Decision Making

Decision makers's preferences play an important role in the decision-making process. Based on complex scenarios, according to their own knowledge structure, the speculative process of analyzing and dealing with alternatives of decision makers directly affects the group decision-making process. Furthermore, their preferences also affect the alternative selection. How to effectively gather various types of preference information needs to be addressed urgently. According to the structure categories of preference information given by group decision makers in the decision-making process, the aggregation methods of individual preferences can be generally divided into the same structural preference assembly and different structural preference assembly. The assembly methods of the same structural preference are abundant [47–53], and the aggregation methods of certainty preference with different preference structure have also attracted academic attention. Chiclana and Herrera [54] studied an aggregation method of three forms of preference including utility value, order value, and the complementary judgment matrix in group decision making. Delgado et al. [55] studied an aggregation method of two forms of preference including linguistic judgment matrices and value judgment matrices. Fan and Jiang [56] and Xiao et al. [57] studied an aggregation method of four forms of preference including complementary judgment matrices, reciprocal judgment matrices, utility value, and order value. But in the group preference aggregation studies of the existing literature, the impact of complexity in group decision-making processes was not great. In fact, due to the presence of external disturbances, the complexity of objective issues, the ambiguity of the human mind, and the uncertain environment, using deterministic preference to characterize complex decision problems is unrealistic. Research [58–62] has summarized the uncertain decision-making progress, focusing on problems with single decision-making preference and data structure. In group decision-making processes, aggregation of the uncertain pluralistic structural preference has the following five aspects of difficulties:

1. There is very little research on the complexity of group decision making. There is no effective characterization for the complex structural relationship among decision-making groups (or decision-making alternatives) and no models and methods for the effects internal and external environments effects of on decision-making groups and alternatives. These complex factors will directly affect the assembly model selection and parameter settings.

2. Research on the internal decision-making mechanism of uncertain preference information is not perfect. The traditional decision-making mechanism of the uncertain preference information (such as the interval number reciprocal judgment matrix) has been researched to some extent [63–65] but the weight calculation method of uncertain preference is not perfect. In particulars research on uncertain preference consistency value is lacking, while some new decision-making mechanism of the uncertain preferences is rare [66].

3. Research on the consistent conversion mechanism of various types of uncertain preference information is lacking. The multiple uncertain preference aggregation method can be summarized after aggregating the different structural preferences into the same structure, directly aggregating various forms of preference information without consistent conversion, aggregating based on prioritizing various types of preference information, etc. Information distortion is inevitably produced in the process of aggregating the different structural preferences into the same structure. Mikhailov [67] studied a consistent conversion method of various types of certain preference information, and in Wu [68] Xu [69] several kinds of consistent conversion formulas for various types of uncertain preference information are given. Zhu and Liu [70] studied the changing process of information based on complementary judgment matrices into reciprocal judgment matrices. But there exists no literature researching

equivalence and the validity of information in the consistent conversion process of various types of uncertain preference, which will thus affect the quality and reliability of group preferences aggregation.

4. An optimization model of a variety of uncertain preference aggregation is lacking. Zhu [71] proposed an assembly method for two types of uncertain preferences—interval number complementary judgment matrices and interval number reciprocal judgment matrices. Zhu et al. [72] proposed a measurement of preference information of decision makers utilizing a three-point interval number and studied two aggregating methods of two kinds about the three-point interval number judgment matrices. But there have been no measures to determine experts' weights based on uncertain preference information, how to measure the satisfaction degrees of decision-making groups, how to build an aggregating model with the maximum reflection of decision-making group preferences, and how to aggregate more types of uncertain preferences.

5. An aggregation method of uncertain preference information with timing characteristics is lacking. Researches [73–78] have studied an aggregation method of certain preference data information with timing characteristics, but there has been no literature considering aggregation methods of multiple stages of uncertain group preference information, whose challenges are weighting of decision makers and stages based on uncertain preference information. The challenges are the weight solution of decision makers and stages based on uncertain preference information. It is also important to forecast development trends before making rational decisions.

1.3 Brief Introduction of Content and Chapter

1.3.1 Decision-Making Mechanism of Uncertain Preference Information

The consistency-measuring method of uncertain preference information has been studied, when it would affect the scientific degree of decision-making results directly whether or not the preferences of decision makers were logically consistent. The ranking methods of various types of uncertain preference information have been studied based on the measurement of consistency in order to strengthen the case for scientific decision making. Chapter 2 discusses the consistency-measuring and -ranking methods of interval number reciprocal judgment matrices and interval number complementary judgment matrices, which analyze the nature of the weighting model. Chapter 3 presents a new form of uncertain preference—unascertained number—and defines the consistency for its features, while providing two kinds of sorting methods. Chapter 4 presents a special form of interval number—the three-point interval number—and notes important factors in using such a preference, which is used to streamline an enterprise's paper specification process. Chapter 5 studies the linguistic preference, of which two consistencies have been defined, and an optimization model is established to improve consistency. In addition, aggregating and ranking methods are proposed for a two-tuple linguistic preference.

1.3.2   *Aggregation of Multivariate Uncertain Preference of Decision Makers*

Group satisfaction degree indicators are founded in the uncertain decision context in order to predict and assess the effects of group decision making. Aggregating methods of various types of uncertain preference information are studied to establish the optimal assembly model reflecting the maximum degree of group preference. An aggregation method is proposed based on clustering for the larger decision-making scenarios to enhance the management and control of key decision-making individuals (population) under the uncertain decision-making context and coordinate group differences. Chapter 6 studies aggregating methods of preferences mixed interval number complementary judging information and reciprocal judging information and aggregating methods of four kinds of preferences including utility values, interval number, and order form and also studies large-scale group decision-making methods based on gray clustering. In addition, the chapter proposes aggregating methods of complementary and reciprocal judging information based on modified consistency.

1.3.3 Aggregation of Multivariate Uncertain Preference with Timing Characteristics

Because human cognition and decision making environment development are both time related, comprehensive evaluation must consider not only an alternative's development process but also its potential trends. In the group decision-making process with timing characteristics, consistency differences among various stages of group preferences are analyzed, and an algorithm is researched to look at inconsistencies among various stages of group decision making. Aggregating methods of multistage multiclass preference with timing characteristics are proposed. Thus, the translation methods of different preference are also developed. Chapter 7 studies the multistage aggregating model of complementary and reciprocal judgment matrices based on certainty and uncertainty is suggested and the weight of the stage is studied. Then those methods are applied to the evaluation of a corporate advertising agency. Chapter 8 proposes an aggregating model of multistage linguistic information based on TOPSIS (Technique for Order Preference by Similarity to an Ideal Solution).

The chapter structure of the book can be seen in Figure 1.1.

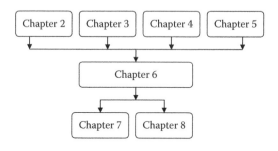

Figure 1.1 Chapter structure.

2

DECISION-MAKING METHOD OF INTERVAL NUMBER RECIPROCAL AND COMPLEMENTARY COMPARISON MATRIX

2.1 Introduction

The analytical hierarchy process (AHP) is widely used in the multiple-criteria decision-making fields (see [79–82]). When using the AHP, a decision maker often makes imprecise judgments or inconsistent judgments due to the complexity of the decision-making problem and adopts the interval numbers for imprecise judgments. Then, an interval number comparison matrix (INCM) is obtained. The INCM has been studied by many researchers, and the main content and contribution of this chapter are as follows:

1. Consistency definition and consistency test approach: To our knowledge, few researchers have focused on the consistency definition. Wei et al. [83] and Bryson and Joseph [84] gave a perfect consistency definition. However, the perfectly consistent comparison matrix cannot be obtained easily. Moreover, the consistency is important and has many effects on the weight estimation. Therefore, a satisfactory consistency definition and its properties should be studied. In addition, there are not practicable approaches to test whether an INCM is consistent.

2. Weight estimation analysis: There are many available approaches to derive weights. However, they have many limitations. For example, the approach from Wei et al. [83] is not obvious and cannot estimate the exact range. Leung

and Cao [85] and Haines [86] put forward a linear model based on the feasible region of a weight. However, it cannot guarantee all vertices of a feasible region found when the rank of the INCM is big enough. Wang et al. [87] did not test whether the matrix was consistent, and the computational work was extensive. Lipovetsky and Tishler [88] derived the weights from the matrix comprised of the stochastic variables, though it did not have reciprocal properties. Byeong [89] derived the weights via a simulation approach. However, the method required comparative computations, in limited iterations, and cannot guarantee the exact range found. Mikhailov [90] and Buckley et al. [91] derived the weights from the fuzzy numbers comparison matrix, though they did not consider the consistency effects on the weight estimation. In this case, the decision-making reliability based on the weights cannot be guaranteed. In order to overcome these limitations, a new weight model solved by the genetic algorithm is put forward. The aim of this chapter is to make the AHP theory [92] perfect under the uncertainty case [98].

2.2 Consistency and Weight Model on Interval Number Reciprocal Comparison Matrix

2.2.1 Consistency Analysis

When studying the weights estimation approach for the fuzzy numbers comparison matrix, Leung and Cao [85] defined the weights feasible region as

$$S = \left\{ w_i \middle| a_{ij}^{L} \le \frac{w_i}{w_j} \le a_{ij}^{U}, w_i > 0, \quad i, j = 1, \ldots, n \right\}.$$

Based on the feasible region, the local consistency and the local satisfactory consistency are defined.

Definition 2.1

Let $j = \{1, \ldots, n\}$, INCM $\bar{A}$ has local consistency if one set of weights $w_i, i = 1, \ldots, n$, meets

$$a_{ij}^{L} \leq \frac{w_i}{w_j} \leq a_{ij}^{U}, \forall i, j \in J \tag{2.1}$$

It should be noted that the reason that the INCM satisfied Formula (2.1) is called *local consistency* instead of consistency, which although there is one set of weights w_i, $i = 1, \ldots , n$, is derived from the matrix A

$$(a_{ij}^{L} \leq a_{ij} \leq a_{ij}^{U}).$$

The weights derived from any one matrix A

$$(a_{ij}^{L} \leq a_{ij} \leq a_{ij}^{U})$$

may not satisfy Formula (2.1). Therefore, Formula (2.1) can represent only the local consistency.

Due to the complexity of the decision-making problem, Formula (2.1) cannot be easily satisfied. In this case, the local satisfactory consistency is defined.

Definition 2.2

The INCM $\bar{A}$ has local satisfactory consistency if one set of weights w_i, $i = 1, \ldots , n$, meets

$$\left(1 - \delta_{ij}\right) a_{ij}^{L} \leq \frac{w_i}{w_j} \leq a_{ij}^{U} \left(1 + \delta_{ij}\right), \forall i, j \in J \tag{2.2}$$

In (2.2), δ_{ij} is the tolerance deviation. Let $\delta = \max_{i,j} \in \{1, \cdots, n\}$. From Xu [80],

$$\lambda_{\max} - n \leq \frac{(n-1)}{2} \delta^2$$

can be obtained, which can be transformed into

$$\frac{\lambda_{\max} - n}{(n-1)RI} \leq \frac{\delta^2}{2RI}.$$

If it has the satisfactory consistency, the formula

$$\frac{\lambda_{max} - n}{(n-1)RI} \le 0.1$$

can be obtained. Then, one can obtain

$$\frac{\delta^2}{2RI} \le 0.1.$$

Therefore, we can obtain $\delta = \sqrt{0.2\,RI}$.

The weights set from Formula (2.2) is also considered the feasible region. Hence, the problem of testing whether the INCM $\bar{A}$ is consistent can be transformed into testing whether S is empty. Based on this idea, model $P_{2.1}$ is put forward.

$$\min \beta = \beta_1 + \beta_2 \qquad (2.3)$$

The following formulas are subject to restrictive conditions:

$$\ln(1-\delta)a_{ij}^{\ L} \le \ln w_i - \ln w_j + \beta_{1ij}, 1 \le i < j \le n \qquad (2.4)$$

$$\ln w_i - \ln w_j \le \ln(1+\delta)a_{ij}^{\ U} + \beta_{2ij}, 1 \le i < j \le n \qquad (2.5)$$

$$\beta_1 \ge \beta_{1ij}, \beta_2 \ge \beta_{2ij} \qquad (2.6)$$

$$\beta_{1ij}, \beta_{2ij} \ge 0, w_i > 0 \qquad (2.7)$$

Formula (2.3) denotes minimizing all tolerance deviations; (2.4) and (2.5) are obtained from the natural logarithm of (2.2); (2.6) denotes β_1, β_2 the maximum of the tolerance deviation; (2.7) denotes all tolerance deviations, and the weights are not negative.

Let $w_i' = \ln w_i$. Model $P_{2.1}$ can be transformed into a linear program. If $\beta = 0$, it means the weights feasible region S is not empty with δ, and the INCM $\bar{A}$ has the local satisfactory consistency; else (min $\beta > 0$) means that the INCM $\bar{A}$ does not have local satisfactory consistency.

How does one test total consistency if the INCM has satisfactory local consistency? When using AHP, there should be one satisfactory consistent matrix $A = (a_{ij})_{n \times n}$, $a_{ij} \in \overline{a_{ij}}$, which can reflect the decision maker's actual preference. However, it cannot be obtained easily because the decision-making problem is complex. To overcome this difficulty, one can generate a satisfactory consistent comparison matrix to denote one's preference approximately. Therefore, the random crisp comparison matrix is introduced.

Definition 2.3

The comparison matrix $A = (a_{ij})_{n \times n}$ is called the *random crisp comparison matrix* of the INCM $\overline{A}$, which is constructed as follows: for $1 \leq i < j \leq n$, a_{ji}, is generated by the uniform distribution in $[a_{ij}^L, a_{ij}^U]$. Let $a_{ji} = 1/a_{ij}$ for $1 \leq i < j \leq n$, and let $a_{ii} = 1$ for all $i = 1, \ldots, n$.

The relationship between the random crisp comparison matrix based on Definition 2.3 and the weights feasible region S from model $P_{2.1}$ can fit Theorem 2.1.

Theorem 2.1

The solution in the weights feasible region corresponds to the satisfactory random crisp comparison matrix uniquely.

Proof

① the comparison matrix $A = (a_{ij})_{n \times n}$ can be constructed via the comparison based on w_i if it satisfies

$$a_{ij}^L \leq \frac{w_i}{w_j} \leq a_{ij}^U.$$

Then, one can obtain

$$a_{ij} = \frac{w_i}{w_j}, a_{ik} = \frac{w_i}{w_k}, a_{kj} = \frac{w_k}{w_j}.$$

That is, $a_{ij} = a_{ik}a_{kj}$, which means the comparison matrix A is perfectly consistent.

② If there is one random crisp comparison matrix having perfect consistency, one can obtain $a_{ij}^L \le a_{ij} \le a_{ij}^U$ based on Definition 2.3. In addition, the formula

$$a_{ij} = \frac{w_i}{w_j}$$

can be obtained. Therefore, one can obtain

$$a_{ij}^L \le \frac{w_i}{w_j} \le a_{ij}^U.$$

If the random crisp comparison matrix A meets $a_{ij}^L(1-\delta) \le (w_i/w_j) \le a_{ij}^L$, $a_{ij}^U \le (w_i/w_j) \le a_{ij}^U(1+\delta)$. From ①, one can conclude that it does not have perfect consistency because the matrix A does not meet

$$a_{ij}^L \le \frac{w_i}{w_j} \le a_{ij}^U.$$

However, it has satisfactory consistency according to the value of δ.

To one random crisp comparison matrix A, if it has satisfactory consistency, from ② one can conclude

$$a_{ij}^L(1-\delta) \le \frac{w_i}{w_j} \le a_{ij}^L \quad \text{or} \quad a_{ij}^U(1+\delta) \ge \frac{w_i}{w_j} \ge a_{ij}^U,$$

which meets

$$a_{ij}^L(1-\delta) \le \frac{w_i}{w_j} \le a_{ij}^U(1+\delta).$$

Via the AHP theory, the weights $w_i = 1, \dots, n$ exist uniquely. Therefore, we can conclude that the solution in the feasible region corresponds to the satisfactory random crisp matrix uniquely.

If the weights feasible region S is empty, there is no satisfactory crisp comparison matrix of the INCM. If S has one set of weights, it must correspond to one satisfactory random crisp comparison matrix A uniquely. Moreover, the larger the number of solutions in the feasible region, the larger the number of satisfactory crisp comparison matrices. In other words, the number of satisfactory random crisp comparison matrices denoting the decision maker's preference is larger. In this case, the decision maker's judgment is right and the logical harmonization is high. Therefore, the number of solutions in the feasible region can test the whole consistency. However, it is difficult to solve the weights feasible region from $P_{2.1}$ because the rank of the INCM is large. To avoid this difficulty, an equivalent approach is put forward.

Definition 2.4

Generate N random crisp comparison matrices based on Definition 2.3, denoted as $A^1, \ldots, A^N$. If there are m comparison matrices having satisfactory consistency ($CR \leq 0.1$), the consistency degree η of the INCM can be defined as

$$\frac{m}{N} \times 100\%.$$

Since the satisfactory random crisp comparison matrix corresponds to the solutions in the weights feasible region uniquely, the value of η can reflect the consistency. In addition, the better the consistency is, the larger value of η is and vice versa. Two examples from previous literature show the consistency degree of Definition 2.4.

Example 2.1:

$$\bar{A} = \begin{vmatrix} [1,1] & [2,4] & [3,5] & [3,5] \\ [1/4,1/2] & [1,1] & [1/2,1] & [2,5] \\ [1/5,1/3] & [1,2] & [1,1] & [1/3,1] \\ [1/5,1/3] & [1/5,1/2] & [1,3] & [1,1] \end{vmatrix}.$$

Generate 50 random crisp comparison matrices and obtain $\eta = 33\%$. This shows that the degree of consistency is poor.

Example 2.2:

$$\bar{A} = \begin{vmatrix} [1,1] & [2,5] & [2,4] & [1,3] \\ [1/5,1/2] & [1,1] & [1,3] & [1,2] \\ [1/4,1/2] & [1/3,1] & [1,1] & [1/2,1] \\ [1/3,1] & [1/2,1] & [1,2] & [1,1] \end{vmatrix}.$$

The value of η equals 88%, which shows that the degree of consistency is high.

Therefore, one can calculate the degree of consistency instead of solving the weights feasible region when analyzing the consistency of the INCM.

2.2.2 Weight Model of Interval Number Comparison Matrix

2.2.2.1 Weight Range Model Let $IN = \{1, \dots, N\}$. Generate N random crisp comparison matrices based on Definition 2.3, denoted as A^I, $a_{ij}^I \in \bar{a}_{ij}$, $I \in IN$. Solve the eigenvalue problem of A^I if $CR(A^I) \leq 0.1$ and obtain the standardization weights denoted as $w^I = (w_1^I, \dots, w_n^I)^T$. If the value of N is large enough, the lower weight range of the INCM $\bar{A}$ can be written as

$$P_{2.2} \quad w_i^L = \min_{I \in IN} w_i^I \tag{2.8}$$

And the upper weight range can be written as

$$P_{2.3} \quad w_i^U = \max_{I \in IN} w_i^I \tag{2.9}$$

From Formulas (2.8) and (2.9), the weights can be expressed as

$$\bar{w}_i = [w_i^L, w_i^U] \tag{2.10}$$

Then, the weights estimation problem can be transformed into estimating values of w_i^L and w_i^U. For solving convenience, (2.8) can be rewritten into (2.11), where G is a constant.

$$P_{2.2} \quad w_i^L = \max_{I \in IN} (G - w_i^I) \tag{2.11}$$

Theorem 2.2

$P_{2.2}$ and $P_{2.3}$ have a feasible solution if the INCM has local satisfactory consistency. In addition, the optimization solution satisfies

$$w_i^L \geq \frac{1}{(n-1)a+1}$$

and

$$w_i^U \leq \frac{1}{\dfrac{n-1}{a}+1},$$

where n denotes the rank of the INCM and a denotes the maximal scale value.

Proof

There will be one random crisp comparison matrix satisfying $CR(A^I) \leq 0.1$ if the INCM has local satisfactory consistency. Hence, $P_{2.2}$, $P_{2.3}$ have a feasible solution.

Let n alternatives be compared. Construct the matrix

$$A' = (a_{ij}')_{n \times n} = \begin{vmatrix} 1 & 1/a & 1/a & \ldots & 1/a \\ a & 1 & 1 & \ldots & 1 \\ a & 1 & 1 & \ldots & 1 \\ \ldots & \ldots & \ldots & \ldots & 1 \\ a & 1 & 1 & 1 & 1 \end{vmatrix}.$$

This shows that w_1' is a minimum among n alternatives. That is, w_1^F from any one consistent comparison matrix A^F is not less than w_1'. The weights from the eigenvector method are close to the geometric mean. Hence, one can obtain

$$w_1' = \frac{\sqrt[n]{\dfrac{1}{a^{n-1}}}}{(n-1)\sqrt[n]{a} + \sqrt[n]{\dfrac{1}{a^{n-1}}}} = \frac{1}{(n-1)a+1}.$$

Therefore, we can obtain

$$w_1^L \geq \frac{1}{(n-1)a+1}.$$

Similarly, construct the reciprocal matrix

$$A'' = (a_{ij}'')_{n \times n} = \begin{vmatrix} 1 & a & a & \cdots & a \\ 1/a & 1 & 1 & \cdots & 1 \\ 1/a & 1 & 1 & \cdots & 1 \\ \cdots & \cdots & \cdots & \cdots & 1 \\ 1/a & 1 & 1 & 1 & 1 \end{vmatrix}.$$

This shows that w_1'' is a maximum among n alternatives. That is, w_1^F from any one consistent comparison matrix A^F is not more than w_1'. One can obtain

$$w_1'' = \frac{\sqrt[n]{a^{n-1}}}{(n-1)\sqrt[n]{1/a} + \sqrt[n]{a^{n-1}}} = \frac{1}{\dfrac{n-1}{a}+1}.$$

as $CR(A'') \leq 0.1$. Therefore, we can obtain

$$w_1^U \leq \frac{1}{\dfrac{n-1}{a}+1}.$$

In a word, $P_{2.2}$ and $P_{2.3}$ have the feasible solution as $CR(A^I) \leq 0.1$. Moreover, we can obtain

$$\frac{1}{\dfrac{n-1}{a}+1} \geq w_i^I \geq \frac{1}{(n-1)a+1}.$$

In particular, $a = 9$ as one adopts a 1–9 scale, with $n = 4$. Then, one can obtain $w_i^L \geq 0.036$ and $w_i^U \leq 0.75$.

2.2.2.2 Possible Value Model The consistency level contained in the comparison matrix can reflect the logical harmonization of the decision maker. One could conclude that the lower CR is, the better the reliability of the weights is if the decision maker does not have a prejudice for any particular alternative. Based on this conclusion, model $P_{2.4}$ was developed. If there are many solutions in the feasible region, they must correspond to satisfactory random crisp comparison matrices uniquely. Among these matrices, there will be a comparison matrix A^g of which the CR is a minimum. Moreover, A^g is unique as long as the INCM is given reasonably. Since the decision-making reliability is higher based on the weights derived from A^g, one can take the weights from A^g as the most possible weights. Then, the weights estimation problem can be transformed into determining A^g. Generate N random crisp comparison matrices based on Definition 2.3, denoted as $A^I, I \in IN$. Solve the eigenvalue problem of A^I. Let $CR(A^g) = \min\{CR(A^I)|I \in IN\}$. Then one can obtain A^g through solving model $P_{2.4}$.

$$P_{2.4}: \min\{CR(A^I)|\ I \in IN\} \qquad (2.12)$$

For solving convenience, (2.12) can be transformed into (2.13).

$$\arg\ \max\ \ G - CR(A^I) \qquad (2.13)$$

The property of $P_{2.4}$ can follow Theorem 2.3.

Theorem 2.3
The optimization solution is not less than zero. Moreover, if the INCM has satisfactory consistency, one can conclude that $CR \leq 0.1$.

Proof
Omitted.

If the optimization solution of model $P_{2.4}$ is not less than 0.1, the INCM does not have local satisfactory consistency. In this case, the decision maker should adjust their judgments to ensure the INCM has satisfactory local consistency.

2.2.2.3 Integration Model The weights of the INCM $\bar{A}$ can be expressed as $\bar{w}_i = [w_i^L, w_i^U]$, which depicts the exact weights range. However, the interval range is too large to guarantee the weights range integrality, which increases the uncertainty. Via model $P_{2.4}$, the decision maker can know the most possible weights, but this result does not consider the uncertainty. In this case, if we integrate these results and express the weights as $\bar{w}_i = [w_i{}^L, w_i{}^g, w_i{}^U]$, the integration result will have the merits of models $P_{2.2}$, $P_{2.3}$, and $P_{2.4}$.

Due to the similarity of $[w_i{}^L, w_i{}^g, w_i{}^U]$ and a triangle fuzzy number, one can adopt its arithmetic approach to obtain the synthesis weights result. Then, rank these interval numbers according to the approach from Chen [92], simplified as

$$y(w_i) = \frac{w_i{}^L + 4w_i{}^g + w_i{}^U}{6}, i = 1, \ldots, n \qquad (2.14)$$

Based on the value of $y(w_i)$, one can obtain the rank of n alternatives.

2.2.3 Algorithm Research

Since the relationship between $CR(A^I)$ and A^I is nonlinear, models $P_{2.2}$, $P_{2.3}$, $P_{2.4}$ are nonlinear programs. If adopting the general approach, the arithmetic is complex. In this chapter, the genetic algorithm is designed. Define parameters as follows. N denotes the individual population number and k denotes the iterations. X denotes the chromosome. $f(X)$ denotes the fitness. Let $\bar{A} = ([a_{ij}{}^L, a_{ij}{}^U])_{n \times n}$. The genetic algorithm is designed as follows:

1. Code chromosome: Let $X = (a_{12}, a_{13}, \ldots, a_{ij}, \ldots, a_{n-1n})$, $1 \leq i < j \leq n$. The value of a_{ij} is randomly generated by the uniform distribution in $[a_{ij}{}^L, a_{ij}{}^U]$. In addition, one gene corresponds to one entry of the upper triangular matrix and one chromosome corresponds to one random crisp comparison matrix uniquely.
2. Fitness function: Let $f(X^I) = G - w_i{}^I - b$ denote the fitness of chromosome I as $CR(A^I) > 0.1$, where b denotes the nonfeasible punishment factor. Generally, it is close to 1. Let $f(X^I) = G - w_i{}^I$ as $CR(A^I) \leq 0.1$ as. In model $P_{2.4}$, the fitness is $f(X^I) = G - CR(A^I)$.
3. Crossover operator: The arithmetic crossover is adopted.
4. Mutation operator: The uniform mutation operator is adopted.
5. Selection operator: Models $P_{2.2}$, $P_{2.3}$ are to solve the standardization weights and $P_{2.4}$ is to solve the consistency ratio. The

fitnesses of models $P_{2.2}$, $P_{2.3}$, and $P_{2.4}$ are not large. Hence, the wheel selection cannot differentiate these chromosomes. In this case, one can rank finesses of N chromosomes from small to large and let them be the value of 1 to N accordingly. Then, the selection probability can be calculated as

$$p(I) = \frac{2I}{N(N+1)},$$

for all $I = 1, \ldots, N$.

6. Stop rule: The iteration continues until a given number of generations is reached.

The algorithm steps can be described as follows:

Step 1: Generate the initialization population.
Step 2: Solve $f(X)$ and select N populations based on the selection operator. If it satisfies the stop rule, stop; else go to step 3.
Step 3: Generate N new populations via the crossover and mutation operator, then go to step 2.

2.2.4 Roll Vendor Evaluation Using Interval Number Method

2.2.4.1 Review of Vendor Selection Approaches Logistics management is the third source of benefit of an enterprise as a new research topic following the production and sale, which plays an important role in market competition. Purchasing logistics as a key component of an enterprise logistics system takes an important proportion. It has a significant effect on normal production and economic benefit. In addition, it is a crucial part of the production of an enterprise. The steel-iron industry is representative of process manufacturing. The purchasing costs of raw materials, machine equipment, and spare parts hold a dominant fraction of production cost. The characteristics of the steel-iron enterprise are a continuity of the production procedure, the dispersion of vendors, and strict requirements for quality, which lead to the purchasing problem being more complex and important. After becoming a member of World Trade Organization (WTO), the steel-iron enterprises in China were challenged by multinational corporations, which require optimization of not only the interior logistics but also the exterior logistics. In this new situation, the relationship between the purchaser and vendor is changed greatly. The purchase action is not only involved in cost control but also affects the morality and credit standing of the enterprises. The purchasing management level becomes the core competition of a large-scale steel-iron enterprise. Due to the purchasing particularities of steel-iron, the related literature for a steel-iron enterprise is lacking. Unfortunately, no literature has focused on the roll purchasing problem. The roll is a crucial spare part of the enterprise production, which has properties of high volume and high cost. As a result, its purchasing cost and quality directly affect an enterprise's competition. In many large-scale steel-iron enterprises, the roll vendor selection and roll production management have been considered in enterprise resource planning (ERP).

Vendor selection is a complex decision-making problem since many criteria should be considered. Some criteria are incompatible with other criteria. Moreover, some criteria are quantitative and some criteria are qualitative. Many authors have studied the selection problem. Kaslingam and Lee [93] proposed a mixed integer programming model for selecting vendors and determining order quantities. Degraeve et al. [94] compared the relative efficiency of selection models having been published in the purchasing literature according to the concept of total cost of ownership. These authors formulated vendor selection as a multiple-criteria decision-making problem. However, they considered only quantitative criteria, and none took qualitative criteria into account. Some authors formulated vendor selection as a multiple-criteria decision-making problem and considered the quantitative and qualitative criteria for these processes. Talluri et al. [95] proposed a two-phase quantitative framework to aid the decision-making process in selecting a compatible set of partners. However, the shortcoming is that the model cannot adequately represent the specific features of the selection problem. Some authors combined the AHP and mathematical programming for the vendor selection process. Masood [96] identified five sets of quality measures measured by the AHP to select the best set of quality control instruments. Then a goal-programming model was incorporated. Ghodsypour and Brien [97] integrated the AHP and linear programming to consider both tangible and intangible factors in choosing the best suppliers and place the optimum quantities among them such that the total value of purchasing became maximal. Most of these approaches consider only quantitative criteria.

The roll vendor selection process should be considered as a multiple criteria decision-making problem, rather than a pure mathematical modeling problem. The AHP is fit for vendor selection due to its inherent capability to handle the qualitative and quantitative criteria, its simple and understandable decision procedure, and the effective evaluation and selection process. In this chapter, the AHP is adopted for a large-scale steel-iron enterprise in China. Many authors used the original AHP, where the pairwise comparison judgments are represented as exact numbers. However, in many practical cases, the human preference model is uncertain, and the decision maker is reluctant or unable to assign exact numerical values to the comparison ratios. In order to deal with such an uncertain evaluation problem, a modification of the AHP method based on the interval number comparison judgments was proposed by Saaty and Vargas [98]. The main approach for the roll vendor selection in this chapter is based on the interval number comparison judgments. A new approach to derive the weight from the interval number comparison judgments is put forward.

2.2.4.2 Criteria for the Evaluation Problems Enterprise A is one of the largest steel-iron enterprises in China. It can produce 20 million tons of steel per annum. Steel production is concentrated in the automobile, oil, natural gas, and other heavy industries. Consumption of steel rolls totals 200 million yuan annually. If the roll vendor is reasonably selected, it can save much production cost and can increase the core competition of the enterprise to a great degree. The operating manager has to select one roll vendor supplying one kind of roll used in the finishing production line, where it has seven working machines, F1 to F7. The related technique details are not covered in this chapter.

According to experts in the field and the principles of index systems, the comprehensive evaluation index system of rollers is established, shown in Figure 2.1. These indices are explained as follows:

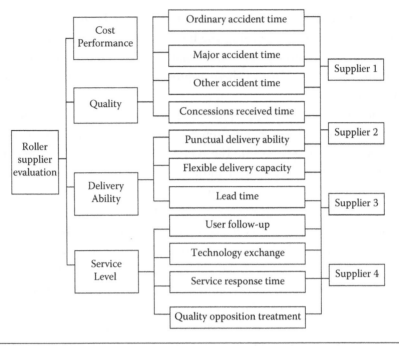

Figure 2.1 Hierarchy of alternatives and criteria for roll vendors.

1. Cost performance: a composite index reflecting the roller performance. Suppose that the price of a roller steel is p, and the total length of its life is q. Considering that the difficulties of rolled products vary, the actual calculation of the total length of rolling is $q' = q \times k$, where k is the correction factor (given by experts); then the cost performance of the supplier's roll is defined as p/q'. The smaller the value is, the better the overall performance of the roller is and vice versa.

2. Quality: measured through the number of the accidents and concessions. According to the influence on production, the character of an accident can be classified as ordinary, major, and other. The accident does not cause discontinued production lines, and major economic loss is called an *ordinary accident*. One that results in production line shutdown and major economic losses is called a *major accident;* an *other accident* is one that seriously affects the quality of products, although no production lines are discontinued, and results in larger

economic loss; *concessions received* refers to a new roller in the delivery not fully meeting contract requirements for quality and technical standards but still received.

3. Delivery ability: evaluated through the suppliers' punctual delivery ability, flexible delivery ability, and ordering lead time. A delivery date earlier than the date stipulated in the contract will result in additional costs of inventory management and increase storage space requirements, while the tardiness of delivery may affect normal production. Flexible delivery reflects the supplier's ability for emergency orders, which is an important indicator to measure the delivery ability. *Lead time* is the time needed by a supplier to deliver a component when needed. It has a great impact on procurement strategies.

4. Service level: evaluated through user follow-up, technology exchange, service response time, and ability of a supplier to meet quality needs.

Subsequent parts of this chapter evaluate a roller supplier based on the index system as shown in Figure 2.1 to indicate the application of this method for the evaluation process. Suppose four suppliers are denoted 1 through 4.

2.2.4.3 Evaluation Result and Analysis

1. Criteria weights: judge the four criteria and obtain an interval number comparison matrix shown in Table 2.1. According to the method above, the result is shown as rows 6 and 7, below, where $[w_i^L, w_i^U]$ denote the lower and upper weight limits of a certain criteria or program, and w_i^g is the most likely value of the weight. Table 2.1 shows that cost performance is the most important because of its direct impact on the level of roller consumption.

2. Subcriteria weights.
   ① The comparison matrix of suppliers based on cost performance criteria as shown in Table 2.2.
   ② Comparison of suppliers based on quality criteria: give each supplier 100 points. According to the supplier's unusual use of roller, the severity of the accident is considered, and then

Table 2.1 Comparison Matrix and Weights of the Criteria

OBJECT	COST PERFORMANCE	QUALITY	DELIVERY ABILITY	SERVICE LEVEL	$[w_i^L, w_i^U]$	w_i^g
Cost performance	[1, 1]	[2, 4]	[3, 5]	[3, 5]	[0.452, 0.565]	0.532
Quality	[1/4, 1/2]	[1, 1]	[1/2, 1]	[2, 5]	[0.172, 0.256]	0.198
Delivery ability	[1/5, 1/3]	[1, 2]	[1, 1]	[1/3, 1]	[0.115, 0.197]	0.143
Service level	[1/5, 1/3]	[1/5, 1/2]	[1, 3]	[1, 1]	[0.086, 0.163]	0.128

Table 2.2 Comparison Matrix and Weights of Cost Performance Criteria

COST PERFORMANCE	SUPPLIER 1	SUPPLIER 2	SUPPLIER 3	SUPPLIER 4	$[w_i^L, w_i^U]$	w_i^g
Supplier 1	[1, 1]	[1, 2]	[1/3, 1/2]	[1/4, 1/2]	[0.111, 0.182]	0.138
Supplier 2	[1/2, 1]	[1, 1]	[1/4, 1/3]	[1/5, 1/4]	[0.083, 0.111]	0.097
Supplier 3	[2, 3]	[3, 4]	[1, 1]	[1/2, 1]	[0.285, 0.387]	0.334
Supplier 4	[2, 4]	[4, 5]	[1, 2]	[1, 1]	[0.363, 0.487]	0.431

Table 2.3 Roller Vendor Score of Quality Criteria

QUALITY	ORDINARY ACCIDENT TIME	MAJOR ACCIDENT TIME	OTHER ACCIDENT TIME	CONCESSIONS RECEIVED TIME	SCORE
Supplier 1	1	0	4	0	55
Supplier 2	2	1	0	3	45
Supplier 3	6	0	3	1	30
Supplier 4	1	2	0	4	25

deducts some score once each: 5 points will be deducted once an ordinary accident occurs, 15 for points a major accident, 10 points for other accident, and 10 points for concessions received. According to the use histories of rollers, the four suppliers' scores are shown in Table 2.3.

Translate the supplier's score into the weight form of AHP: obtain the supplier's percentile score under a scoring rule denoted $r_1, \ldots, r_n$. Suppose that $r^L = \min_i r_i$, $r^U = \max_i r_i$, the importance ratio of r^U and r^L given by decision makers is d, and

$$r_{ij} = \frac{r_i - r_j}{r^U - r^L} \times d,$$

so structure matrix $B = (b_{ij})_{n \times n}$ can be built such that

$$b_{ij} = \begin{cases} r_{ij}, & r_{ij} > 1 \\ 1, & |r_{ij}| \leq 1 \\ -1/r_{ij}, & r_{ij} < -1 \end{cases}.$$

Weights of the suppliers are obtained based on comparison matrix B. In this case, based on quality criteria, the ratio of the best to worst suppliers is 4 to 1 by means of AHP. The comparison matrix based on quality criteria is obtained as shown in Table 2.4. As $CR = 0.012$, the comparison matrix is of acceptable consistency. The weights are shown in Table 2.4 in the rightmost column.

③ Comparison matrices of subcriteria based on delivery ability and suppliers based on subcriteria are shown in Tables 2.5 to 2.8. Since order management and production

Table 2.4 Comparison Matrix and Weights of Quality Criteria

QUALITY	SUPPLIER 1	SUPPLIER 2	SUPPLIER 3	SUPPLIER 4	WEIGHT
Supplier 1	1	4/3	10/3	4	0.442
Supplier 2	3/4	1	2	8/3	0.305
Supplier 3	3/10	1/2	1	1	0.134
Supplier 4	1/4	3/8	1	1	0.119

Table 2.5 Comparison Matrix and Weights of Delivery Ability Criteria

DELIVERY ABILITY	FLEXIBLE DELIVERY	PUNCTUAL DELIVERY	ORDERING LEAD TIME	$[w_i^L, w_i^U]$	w_i^g
Flexible delivery	[1, 1]	[1/4, 1/3]	[1, 2]	[0.167, 0.204]	0.186
Punctual delivery	[3, 4]	[1, 1]	[2, 4]	[0.615, 0.642]	0.630
Ordering lead time	[1/2, 1]	[1/4, 1/2]	[1, 1]	[0.160, 0.194]	0.184

Table 2.6 Comparison Matrix and Weights of Flexible Delivery Ability Criteria

FLEXIBLE DELIVERY	SUPPLIER 1	SUPPLIER 2	SUPPLIER 3	SUPPLIER 4	$[w_i^L, w_i^U]$	w_i^g
Supplier 1	[1, 1]	[2, 5]	[2, 4]	[1, 3]	[0.369, 0.537]	0.484
Supplier 2	[1/5, 1/2]	[1, 1]	[1, 3]	[1, 2]	[0.166, 0.272]	0.209
Supplier 3	[1/4, 1/2]	[1/3, 1]	[1, 1]	[1/2, 1]	[0.102, 0.178]	0.137
Supplier 4	[1/3, 1]	[1/2, 1]	[1, 2]	[1, 1]	[0.143, 0.245]	0.169

Table 2.7   Comparison Matrix and Weights of Punctual Delivery Ability Criteria

PUNCTUAL DELIVERY	SUPPLIER 1	SUPPLIER 2	SUPPLIER 3	SUPPLIER 4	$[w_i^L, w_i^U]$	w_i^g
Supplier 1	[1, 1]	[2, 3]	[3, 5]	[4, 5]	[0.467, 0.586]	0.499
Supplier 2	[1/3, 1/2]	[1, 1]	[1, 2]	[3, 4]	[0.187, 0.284]	0.250
Supplier 3	[1/5, 1/3]	[1/2, 1]	[1, 1]	[2, 4]	[0.129, 0.211]	0.167
Supplier 4	[1/5, 1/4]	[1/4, 1/3]	[1/4, 1/2]	[1, 1]	[0.058, 0.092]	0.083

Table 2.8   Comparison Matrix and Weights of Ordering Lead Time Criteria

ORDERING LEAD TIME	SUPPLIER 1	SUPPLIER 2	SUPPLIER 3	SUPPLIER 4	$[w_i^L, w_i^U]$	w_i^g
Supplier 1	[1, 1]	[1, 2]	[5, 6]	[1, 5/4]	[0.367, 0.397]	0.390
Supplier 2	[1/2, 1]	[1, 1]	[3, 4]	[3/5, 1]	[0.234, 0.285]	0.267
Supplier 3	[1/6, 1/5]	[1/4, 1/3]	[1, 1]	[1/4, 1/2]	[0.071, 0.081]	0.076
Supplier 4	[4/5, 1]	[1, 5/3]	[2, 4]	[1, 1]	[0.267, 0.307]	0.287

Table 2.9   Comparison Matrix and Weights of Service-Level Criteria

SERVICE LEVEL	USER FOLLOW-UP	TECHNOLOGY EXCHANGE	SERVICE RESPONSE	QUALITY OPPOSITION TREATMENT	$[w_i^L, w_i^U]$	w_i^g
User follow-up	[1, 1]	[0.22, 0.4]	[2.5, 4]	[3.5, 4.8]	[0.217, 0.317]	0.274
Technology exchange	[2.5, 4.5]	[1, 1]	[3, 4.5]	[4, 6]	[0.479, 0.591]	0.516
Service response	[0.25, 0.4]	[0.22, 0.33]	[1, 1]	[3, 5]	[0.117, 0.171]	0.146
Quality opposition treatment	[0.21, 0.29]	[0.17, 0.25]	[0.2, 0.33]	[1, 1]	[0.055, 0.073]	0.064

planning of the enterprise are reasonable, roller consumption can be measured accurately to some extent. The enterprise requires greater punctual delivery ability but lower flexible delivery ability.

④ The comparison matrix of subcriteria based on service level is shown in Table 2.9, and comparison matrices of roller suppliers based on subcriteria are shown in Tables 2.10 to 2.13.

Table 2.10 Comparison Matrix and Weights of User Visit Criteria

USER FOLLOW-UP	SUPPLIER 1	SUPPLIER 2	SUPPLIER 3	SUPPLIER 4	$[w_i^L, w_i^U]$	w_i^g
Supplier 1	[1, 1]	[0.5, 0.7]	[1, 1]	[0.7, 1]	[0.131, 0.206]	0.187
Supplier 2	[1.4, 2]	[1, 1]	[1.4, 1.8]	[1.2, 1.7]	[0.298, 0.371]	0.351
Supplier 3	[1, 1]	[0.6, 0.7]	[1, 1]	[0.7, 1]	[0.179, 0.221]	0.190
Supplier 4	[1, 1.4]	[0.6, 0.8]	[1, 1.4]	[1, 1]	[0.235, 0.371]	0.271

Table 2.11 Comparison Matrix and Weights of Technique Communion Criteria

TECHNOLOGY EXCHANGE	SUPPLIER 1	SUPPLIER 2	SUPPLIER 3	SUPPLIER 4	$[w_i^L, w_i^U]$	w_i^g
Supplier 1	[1, 1]	[3, 6]	[1, 2]	[1/2, 1]	[0.251, 0.345]	0.311
Supplier 2	[1/6, 1/3]	[1, 1]	[1/2, 1]	[1/5, 1/4]	[0.091, 0.110]	0.100
Supplier 3	[1/2, 1]	[1, 2]	[1, 1]	[1/4, 1/2]	[0.149, 0.160]	0.157
Supplier 4	[1, 2]	[4, 5]	[2, 4]	[1, 1]	[0.392, 0.457]	0.432

Table 2.12 Comparison Matrix and Weights of Service Response Criteria

SERVICE RESPONSE	SUPPLIER 1	SUPPLIER 2	SUPPLIER 3	SUPPLIER 4	$[w_i^L, w_i^U]$	w_i^g
Supplier 1	[1, 1]	[3, 4]	[2, 3]	[2/3, 1]	[0.313, 0.354]	0.327
Supplier 2	[1/4, 1/3]	[1, 1]	[1/3, 1/2]	[1/6, 1/4]	[0.078, 0.089]	0.081
Supplier 3	[1/3, 1/2]	[2, 3]	[1, 1]	[1/3, 1/2]	[0.165, 0.179]	0.170
Supplier 4	[1, 3/2]	[4, 6]	[2, 3]	[1, 1]	[0.378, 0.458]	0.423

Table 2.13 Comparison Matrix and Weights of Doubtful Quality Disposal Ability Criteria

QUALITY OPPOSITION TREATMENT	SUPPLIER 1	SUPPLIER 2	SUPPLIER 3	SUPPLIER 4	$[w_i^L, w_i^U]$	w_i^g
Supplier 1	[1, 1]	[2, 4]	[4/3, 2]	[2, 3]	[0.380, 0.467]	0.402
Supplier 2	[1/4, 1/2]	[1, 1]	[1/3, 1]	[1/2, 1]	[0.122, 0.181]	0.178
Supplier 3	[1/2, 3/4]	[1, 3]	[1, 1]	[1, 3/2]	[0.221, 0.289]	0.252
Supplier 4	[1/3, 1/2]	[1, 2]	[2/3, 1]	[1, 1]	[0.167, 0.213]	0.188

3. Comprehensive weight: according to the data of Tables 2.1–2.13, the weight results of various suppliers are shown in Table 2.14, which is obtained by a weight combination method. According to interval number weights sequencing formula,

Table 2.14 Rolls Vendor Weights of Criteria and Global Weights

COST PERFORMANCE	w_i^L	w_i^g	w_i^U
Supplier 1	0.0502	0.0730	0.1030
Supplier 2	0.0380	0.0520	0.0630
Supplier 3	0.1290	0.1780	0.2190
Supplier 4	0.1640	0.2290	0.2750
QUALITY			
Supplier 1	0.0760	0.0880	0.1130
Supplier 2	0.0520	0.0600	0.0781
Supplier 3	0.0230	0.0260	0.0360
Supplier 4	0.0200	0.0240	0.0350
DELIVERY			
Supplier 1	0.0470	0.0681	0.1109
Supplier 2	0.0207	0.0352	0.0577
Supplier 3	0.0124	0.0207	0.0369
Supplier 4	0.0121	0.0137	0.0251
SERVICE			
Supplier 1	0.0177	0.0365	0.0593
Supplier 2	0.0107	0.0219	0.0344
Supplier 3	0.0122	0.0223	0.0353
Supplier 4	0.0226	0.0475	0.0785
TOTAL WEIGHT BASED ON THE TARGET LAYER			
Supplier 1	0.1909	0.2656	0.3862
Supplier 2	0.1214	0.1691	0.2332
Supplier 3	0.1766	0.2470	0.3272
Supplier 4	0.2187	0.3142	0.4136

$$y(w_i) = \frac{w_i^L + 4w_i^g + w_i^U}{6}, i = 1, \ldots, n$$

The comprehensive index values of the four suppliers are 0.273, 0.172, 0.249, 0.315. So, roller supplier 4 has the highest score, followed by supplier 1 and supplier 3; supplier 2 is the worst. Therefore, roller supplier 4 will be used as the priority supplier.

Rollers are a key spare part of the steel industry, because of the high production costs, high consumption rate, and quality of the supplier, which will impact the quality and stability of the steel industry directly. Therefore, the problems of selecting and evaluating a roller supplier are particularly important. The comments of the roller supplier need to take into account performance, price, quality, delivery capability, and service level, a typical multiattribute decision-making problem, including both quantitative attributes and qualitative attributes. This chapter utilizes iron and steel enterprises in China as the research background, and the roller vendor selection based on interval AHP is suggested.

2.3 Consistency and Weight Model on Interval Number Complementary Comparison Matrix

Scale is one measurement to quantify qualitative determination of decisions in AHP. The connotation is to endow quantitative value to each degree of qualitative performance and also to ensure that the quantitative value accords with the relationship of each qualitative degree. Many kinds of scale have been proposed, and scale 0.1–0.9 has aroused more emphasis because of strong psychological foundation and fuzzy math. Then, complementary judgment matrix comes into being [99]. In practice, due to the complexity of the problem, decision makers cannot provide an explicit judgment of their preference. They prefer to provide an interval number and interval number judgment matrix. Many experts pay much attention to interval number complementary judgment matrix, but the research is scant. This chapter puts forward a new weight solution approach based on stochastic simulation. This approach is simple, and is proved by mathematics.

2.3.1 Basis Definition

The definition of scales 0.1–0.9 is shown is Table 2.15.

Definition 2.5

Suppose that matrix $A = (a_{ij})_{n \times n}$ is a complementary judgment matrix, if it meets the condition $a_{ij} + a_{ji} = 1$, $a_{ii} = 0.5$, $a_{ij} \in \{0.1, \dots, 0.9\}$, $i, j = 1, \dots, n$.

Table 2.15 The Definition of Scales 0.1–0.9

INTENSITY OF IMPORTANCE	DEFINITION
0.1	Latter element strongly favor former element
0.3	Latter element obviously favor former element
0.5	Two elements contribute equally to the objective
0.7	Former element obviously favor latter element
0.9	Former element strongly favor latter element
0.2, 0.4, 0.6, 0.8	Used to express intermediate values

Definition 2.6

Suppose complementary judgment matrix $A = (a_{ij})_{n \times n}$, if $a_{ij} = a_{ik} - a_{jk} + 0.5$ to $\forall i, j, k$, matrix A has complete consistency.

Generally, the larger the difference value between a_{ij} and $a_{ik} - a_{jk} + 0.5$ is, the weaker the consistency of the matrix is.

A row sum normalization approach is widely applied in all of the weight solution approaches of complementary judgment matrix. The calculation formula for raw sum normalization is:

$$w_i = \frac{\sum\limits_{j=1}^{n} a_{ij}}{\sum\limits_{i=1}^{n}\sum\limits_{j=1}^{n} a_{ij}}, \quad i = 1, \ldots, n \tag{2.15}$$

If the consistency of matrix A is weak, it can be changed into matrix $R = (r_{ij})_{n \times n}$ with complete consistency through Formula (2.16):

$$r_i = \sum\limits_{k=1}^{n} a_{ik}, \quad r_{ij} = \frac{r_i - r_j}{2(n-1)} + 0.5 \tag{2.16}$$

Then, use a row sum normalization approach to calculate the weight of R, and get the weight of all schemes.

$$w_i = \frac{\sum\limits_{j=1}^{n} r_{ij}}{\sum\limits_{i=1}^{n}\sum\limits_{j=1}^{n} r_{ij}}, \quad i = 1, \ldots, n \tag{2.17}$$

Definition 2.7

Suppose matrix $\overline{A} = (\overline{a_{ij}})_{n \times n}$ is an interval number complementary judgment matrix, then $\overline{a_{ij}} = [a_{ij}^L, a_{ij}^U]$, $a_{ij}^L \leq a_{ij}^U$, $\overline{a_{ji}} = [1-a_{ij}^U, 1-a_{ij}^L]$, $1 \leq i < j \leq n$, $a_{ii} = [0.5, 0.5]$, $i = 1, \ldots, n$.

2.3.2 Weight Model

If the decision maker cannot provide definite judgment to pairwise comparisons, it will be expressed by interval number $\overline{a_{ij}} = [a_{ij}^L, a_{ij}^U]$. So, the judgment of the decision maker preference is in the $\overline{a_{ij}}$, namely, $a_{ij}^L \leq a_{ij} \leq a_{ij}^U$. In the interval number complementary judgment matrix, there is a determined judgment $A = (a_{ij})_{n \times n}$, $a_{ij} \in \overline{a_{ij}}$ to reflect the decision maker's true preference objectively, especially a_{ij}, which does not obey a certain distribution. Due to the complexity of decision problems, matrix A is difficult to fix. The interval number complementary judgment matrix is estimated approximately through any one matrix generated randomly in the interval number matrix. Thus, this chapter first puts forward a random deterministic judgment matrix and studies the weight solution.

Definition 2.8

Suppose matrix $A = (a_{ij})_{n \times n}$ is the random deterministic judgment matrix of interval number complementary judgment matrix $\overline{A}$. A is as follows: when $1 \leq i < j \leq n$, $a_{ij} \in [a_{ij}^L, a_{ij}^U]$ is created by uniform distribution probability, and $a_{ji} = 1 - a_{ij}$; $a_{ii} = 0.5$, $i = 1, \ldots, n$.

Note $IN = \{1, \ldots, N\}$, N is a constant. For interval number complementary judgment matrix $\overline{A}$, N matrices A^I, $I \in IN$ are made by Definition 2.8 randomly and the weight of matrix A^I is calculated and noted as $w^I = (w_1^I, \ldots, w_n^I)^T$. When N is large enough, the weight w of the interval number complementary judgment matrix is estimated by Formula (2.18).

$$\overline{w_i} = [w_i^L, w_i^U] \tag{2.18}$$

Thus, $w_i^L = \min\{w_i^I | I \in IN\}$, $w_i^U = \max\{w_i^I | I \in IN\}$. The weight solution w of the interval number complementary judgment matrix is transformed into the solution of w_i^L and w_i^U.

In a word, this chapter gives the solution of model $P_{2.5}$ and model $P_{2.6}$ to calculate w_i^L and w_i^U.

$$P_{2.5} \; w_i^L = \min \{w_i^I | I \in IN\} \tag{2.19}$$

$$\text{s.t.} \quad a_{ij}^I \in \overline{a_{ij}}, \quad i, j = 1, \ldots, n \tag{2.20}$$

$$w_i^I = \frac{\displaystyle\sum_{j=1}^{n} a_{ij}^I}{\displaystyle\sum_{i=1}^{n}\sum_{j=1}^{n} a_{ij}^I}, \quad i = 1, \ldots, n \tag{2.21}$$

$$\text{or} \quad w_i^I = \frac{\displaystyle\sum_{j=1}^{n} r_{ij}^I}{\displaystyle\sum_{i=1}^{n}\sum_{j=1}^{n} r_{ij}^I}, \quad r_{ij}^I = \frac{r_i^I - r_j^I}{2(n-1)} + 0.5, \quad r_i^I = \sum_{k=1}^{n} a_{ik}^I \tag{2.22}$$

$$P_{2.6} \; w_i^U = \max \{w_i^I | I \in IN\} \tag{2.23}$$

$$\text{subject to} \quad a_{ij}^I \in \overline{a_{ij}}, \quad i, j = 1, \ldots, n \tag{2.24}$$

$$w_i^I = \frac{\displaystyle\sum_{j=1}^{n} a_{ij}^I}{\displaystyle\sum_{i=1}^{n}\sum_{j=1}^{n} a_{ij}^I}, \quad i = 1, \ldots, n \tag{2.25}$$

or

$$w_i^I = \frac{\displaystyle\sum_{j=1}^{n} r_{ij}^I}{\displaystyle\sum_{i=1}^{n}\sum_{j=1}^{n} r_{ij}^I}, \quad r_{ij}^I = \frac{r_i^I - r_j^I}{2(n-1)} + 0.5, \quad r_i^I = \sum_{k=1}^{n} a_{ik}^I \tag{2.26}$$

Formula (2.19) denotes the minimum of weight w_i in N complementary judgment matrices. Formula (2.20) denotes a random deterministic judgment matrix, which meets $a_{ij}^L \leq a_{ij}^I \leq a_{ij}^U$, which is made by an interval number complementary judgment matrix. Formula (2.21) denotes the weight of the random deterministic judgment matrix using row sum normalization. Formula (2.22) denotes that when the consistency of A^I is weak, a weight solution is obtained through a mathematical transform. Formula (2.23) denotes the maximum of weight ponderance w_i in N complementary judgment matrices. Formula (2.24) to Formula (2.26) are the same as in model $P_{2.5}$.

The idea of models $P_{2.5}$ and $P_{2.6}$ is to search a certain matrix A^I from interval number complementary judgment matrix $\overline{A}$, and its weight is the minimum or maximum of weight from N matrices. This is used to estimate the weight distribution of the interval number complementary judgment matrix with explicit meaning. Because N matrices are made randomly, the value of w_i^L, w_i^U is related to the large of N, so this model possesses randomness. Theoretically, the value of N should be $N \to \infty$, but this will require a huge calculation. Next we analyze the characteristic and simplify calculation.

2.3.3 Algorithm Research

Theorem 2.4
To complementary matrix

$$A = (a_{ij})_{n \times n}, \sum_{j=1}^{n} \sum_{i=1}^{n} a_{ij} = \frac{n^2}{2}.$$

Proof
Shown in Xu [99].

Theorem 2.5
For a certain order n interval number complementary judgment matrix, use Formula (2.15) or Formulas (2.16) and (2.17) to calculate weight, so the weight w_i is a minimum of complementary judgment matrix A_i', namely, $w_i^L = w_i' = \min_{I \in IN} w_i^I$. So the weight w_i is a maximum of complementary judgment matrix A_i'', namely, $w_i^U = w_i'' = \max_{I \in IN} w_i^I$, therefore,

$$
A_i' = \begin{vmatrix}
0.5 & * & * & \cdots & * \\
* & 0.5 & * & * & * \\
a_{i1}{}^L & a_{i2}{}^L & a_{i3}{}^L & \cdots & a_{in}{}^L \\
\cdots & * & * & \cdots & * \\
* & * & * & * & 0.5
\end{vmatrix} \tag{2.27}
$$

$$
A_i'' = \begin{vmatrix}
0.5 & * & * & \cdots & * \\
* & 0.5 & * & * & * \\
a_{i1}{}^U & a_{i2}{}^U & a_{i3}{}^U & \cdots & a_{in}{}^U \\
\cdots & * & * & \cdots & * \\
* & * & * & * & 0.5
\end{vmatrix} \tag{2.28}
$$

In Formulas (2.27) and (2.28), the "*" denotes that this element is any value if it meets complementary property $(a_{ij} + a_{ji} = 1)$.

Proof

① If complementary judgment matrix uses Formula (2.15) to solve weight, the denominator of w_i^I is $n^2/2$.

So

$$
\min_{I \in IN} w_i^I = \min_{I \in IN} \frac{\sum\limits_{j=1}^n a_{ij}{}^I}{n^2/2} = \frac{\min\limits_{I \in IN} a_{i1}{}^I + \ldots + \min\limits_{I \in IN} a_{in}{}^I}{n^2/2} = \frac{2}{n^2} \sum_{j=1}^n a_{ij}{}^L,
$$

and Formula (2.27) is obtained.
For

$$
\max_{I \in IN} w_i^I = \max_{I \in IN} \frac{\sum\limits_{j=1}^n a_{ij}{}^I}{n^2/2} = \frac{\max\limits_{I \in IN} a_{i1}{}^I + \cdots + \max\limits_{I \in IN} a_{in}{}^I}{n^2/2} = \frac{2}{n^2} \sum_{j=1}^n a_{ij}{}^U,
$$

formula (2.28) is obtained.

② If the complementary judgment matrix uses Formulas (2.16) and (2.17) to solve weight, it is

$$w_i = \frac{\displaystyle\sum_{j=1}^{n} r_{ij}}{\displaystyle\sum_{i=1}^{n}\sum_{j=1}^{n} r_{ij}}.$$

Because matrix R is a complementary matrix, the denominator of

$$w_i^I \quad \text{is} \quad \frac{n^2}{2}.$$

Besides

$$\sum_{j=1}^{n} r_{ij} = \sum_{j=1}^{n}\left(\frac{\displaystyle\sum_{j=1}^{n} a_{ij} - \sum_{j=1}^{n} a_{jj}}{2(n-1)} + 0.5\right)$$

$$= \frac{\displaystyle\sum_{i=1}^{n}\sum_{j=1}^{n} a_{ij} - \sum_{j=1}^{n}\sum_{j=1}^{n} a_{jj} + n(n-1)}{2(n-1)}$$

$$= \frac{n\displaystyle\sum_{j=1}^{n} a_{ij} - \frac{n^2}{2} + n^2 - n}{2(n-1)}.$$

So

$$w_i = \frac{n\displaystyle\sum_{j=1}^{n} a_{ij} + \frac{n^2}{2} - n}{2(n-1)} \times \frac{2}{n^2} = \frac{\displaystyle\sum_{j=1}^{n} a_{ij} + \frac{n}{2} - 1}{n(n-1)}.$$

Then,

$$\min_{I \in IN} w_i{}^I = \min_{I \in IN} \frac{\sum\limits_{j=1}^{n} a_{ij}{}^I + \dfrac{n}{2} - 1}{n(n-1)} = \frac{\sum\limits_{j=1}^{n} a_{ij}{}^L + \dfrac{n}{2} - 1}{n(n-1)},$$

Formula (2.27) is obtained.

Similarly,

$$w_i{}^U = \max_{I \in IN} w_i{}^I = \max_{I \in IN} \frac{\sum\limits_{j=1}^{n} a_{ik}{}^U + \dfrac{n}{2} - 1}{n(n-1)} = \frac{\sum\limits_{j=1}^{n} a_{ik}{}^U + \dfrac{n}{2} - 1}{n(n-1)},$$

Formula (2.28) is obtained.

According to Theorem 2.5, the weight can be solved simply as $w_i{}^L$, $w_i{}^U$. The above proof shows that the present method is the actual row sum normalization method, which is a good value conclusion.

Theorem 2.6

The optimal solution of models $P_{2.5}$ and $P_{2.6}$ meets $w_i{}^L \geq \xi_1$ and $w_i{}^U \geq \xi_2$. If the complementary judgment matrix uses Formula (2.1) to solve weight,

$$\xi_1 = \frac{n+4}{5n^2}$$

and

$$\xi_2 = \frac{9n-4}{5n^2}.$$

If the complementary judgment matrix use Formulas (2.2) and (2.3) to solve weight,

$$\xi_1 = \frac{3}{5n} \quad \text{and} \quad \xi_2 = \frac{7}{5n}.$$

Proof

Assume there are pairwise comparisons of n alternatives

$$
A_i' = \begin{vmatrix}
0.5 & * & * & * & * \\
* & 0.5 & * & * & * \\
0.1 & 0.1 & 0.5 & \cdots & 0.1 \\
\cdots & * & * & \cdots & * \\
* & * & * & * & 0.5
\end{vmatrix}
$$

is made.

This shows that the normalization weight of scheme i is the minimum and the normalization of discretional matrix A^F is $w_i^F \geq w_i'$ in n schemes.

According to Theorem 2.5, if Formula (2.1) is used to solve weight,

$$
w_i' = \frac{2}{n^2} \sum_{j=1}^{n} a_{ij},
$$

when

$$
\sum_{j=1}^{n} a_{ij} \rightarrow \min, \quad w_i' = \frac{0.5 + 0.1(n-1)}{\dfrac{n^2}{2}} = \frac{n+4}{5n^2}.
$$

If Formulas (2.2) and (2.3) are used to solve weight, when

$$
\sum_{j=1}^{n} a_{ij} \rightarrow \min, \quad w_i' = \frac{0.5 + 0.1(n-1) + \dfrac{n}{2} - 1}{n(n-1)} = \frac{3}{5n}.
$$

Similarly,

$$
A_i'' = \begin{vmatrix}
0.5 & * & * & \cdots & * \\
* & 0.5 & * & * & * \\
0.9 & 0.9 & 0.5 & \cdots & 0.9 \\
\cdots & * & * & \cdots & * \\
* & * & * & * & 0.5
\end{vmatrix}
$$

is made. This shows that the normalization weight of scheme i is maximum and the normalization of discretional matrix A^C is $w_i^C \leq w_i''$. According to Theorem 2.6, if Formula (2.1) is used to solve weight, then

$$\sum_{j=1}^{n} a_{ij} \rightarrow \max, \, w_i'' = \frac{0.5 + 0.9(n-1)}{\dfrac{n^2}{2}} = \frac{9n-4}{5n^2}.$$

If Formula (2.1) is used to solve weight, then

$$\sum_{j=1}^{n} a_{ij} \rightarrow \max, \, w_i'' = \frac{0.5 + 0.9(n-1) + \dfrac{n}{2} - 1}{n(n-1)} = \frac{7}{5n}.$$

Theorem 2.7

According to models $P_{2.5}$, $P_{2.6}$ and the weight distribution range w_i^L, w_i^U, there is $w_i^L \leq w_i^U$.

Proof

From Theorem 2, if Formula (2.1) is used to solve weight,

$$w_i^L = \frac{2}{n^2} \sum_{j=1}^{n} a_{ij}^L.$$

Then, $a_{ij}^L \leq a_{ij}^U$, so

$$w_i^L = \frac{2}{n^2} \sum_{j=1}^{n} a_{ij}^L \leq w_i^U = \frac{2}{n^2} \sum_{j=1}^{n} a_{ij}^U.$$

Many weight solutions of the interval number judgment matrix obtain the results $w_i^L > w_i^U$, which go against the definition of interval number and lead the approach lose efficacy. From Theorem 2.7, the approach in this study ensures validity. So, the weights of matrices A_i^L and A_i^U are expressed through the interval number complementary judgment matrix, i.e., $\overline{w}_i = [w_i^L, w_i^U]$, $i = 1, \dots, n$.

2.3.4 Rank Interval Number Based on Possible Method

This chapter analyzed the weight solution of an interval number complementary judgment based on a single rule. It is necessary to compute combined weight to find the relative weight of the alternative relative to the overall goal, namely, $\overline{w}_i = [w_i{}^L, w_i{}^U]$. So, the combined weight is computed within an algorithm of the interval number. For two interval numbers $a = [a^L, a^U]$ and $b = [b^L, b^U]$,

$$a + b = [a^L + b^L, a^U + b^U] \tag{2.29}$$

$$a \times b = [a^L \times b^L, a^U \times b^U] \tag{2.30}$$

The relative weight is obtained after weight combination calculation and is expressed in interval number $\overline{w}_i = [w_i{}^L, w_i{}^U]$. Therefore, the decision maker obtains the distribution range and the priority of each scheme.

The decision maker may rank the interval number weight further and fix the priority.

Definition 2.9
If $a = [a^L, a^U]$, $b = [b^L, b^U]$, $l_a = a^U - a^L$, $l_b = b^U - b^L$, then

$$p(a \geq b) = \frac{\max\{0, l_a + l_b - \max(b^U - a^L, 0)\}}{l_a + l_b} \tag{2.31}$$

is the possibility degree of $a \geq b$.

Do pairwise comparisons with a set of interval numbers, then the possibility degree is obtained by Definition 2.9, and the possibility degree matrix $P = (p_{ij})_{n \times n}$ is made afterwards. This matrix covers all the possibility degree information about all pairwise comparisons. Thereby, the priority of interval number is changed into the weight solution because the possibility degree matrix P is a complementary judgment matrix. The weight w_i, $i = 1, \ldots, n$ is obtained by Formula (2.1) as possibility degree matrix P is complementary judgment matrix. Then, the priority is fixed by w_i.

2.3.5 Example Analysis

Choose the most appropriate office director among four candidates, and use the evaluation rules of work ability, academic level, interpersonal relationship, and physical condition. Assume pairwise comparison using a scale of 0.1–0.9 based on work ability, then the interval number complementary judgment matrix is obtained.

$$
\bar{A} = \begin{bmatrix}
[0.5, 0.5] & [0.3, 0.5] & [0.6, 0.7] & [0.5, 0.6] \\
[0.5, 0.7] & [0.5, 0.5] & [0.4, 0.7] & [0.5, 0.7] \\
[0.3, 0.4] & [0.3, 0.6] & [0.5, 0.5] & [0.2, 0.5] \\
[0.4, 0.5] & [0.3, 0.5] & [0.5, 0.8] & [0.5, 0.5]
\end{bmatrix}
$$

By $\bar{w}_1 = [w_1{}^L, w_1{}^U]$; for instance,

$$
A_1{}^L = \begin{bmatrix}
0.5 & 0.3 & 0.6 & 0.5 \\
* & 0.5 & * & * \\
* & * & 0.5 & * \\
* & * & * & 0.5
\end{bmatrix}, A_1{}^U = \begin{bmatrix}
0.5 & 0.5 & 0.7 & 0.6 \\
* & 0.5 & * & * \\
* & * & 0.5 & * \\
* & * & * & 0.5
\end{bmatrix}.
$$

Using Formula (2.15) to solve weight,

$$
w_1{}^L = \frac{\sum\limits_{j=1}^{n} a_{1j}{}^L}{\sum\limits_{i=1}^{n}\sum\limits_{j=1}^{n} a_{ij}{}^L} = \frac{1.9}{8} = 0.2375, w_1{}^U = \frac{\sum\limits_{j=1}^{n} a_{1j}{}^U}{\sum\limits_{i=1}^{n}\sum\limits_{j=1}^{n} a_{ij}{}^U} = \frac{2.3}{8} = 0.2875,
$$

namely, $\bar{w}_1 = [0.2375, 0.2875]$. Similarly, $\bar{w}_2 = [0.2375, 0.325]$, $\bar{w}_3 = [0.1625, 0.25]$, $\bar{w}_4 = [0.2125, 0.2875]$. By an interval number ranking formula, the priority of all candidates is $w_1 = 0.297$, $w_2 = 0.345$, $w_3 = 0.11125$, $w_4 = 0.246$. It will be seen from this that candidate 2 is the best, followed by candidate 1 and candidate 4, and candidate 3 is the worst. If Formulas (2.16) and (2.17) are used,

$$
w_1{}^L = \frac{1.9 + \dfrac{4}{2} - 1}{4(4-1)} = 0.242,
$$

$$w_1^U = \frac{2.3 + \frac{4}{2} - 1}{4(4-1)} = 0.275,$$

namely, $\overline{w}_1 = [0.242, 0.275]$. Similarly, $\overline{w}_2 = [0.242, 0.3]$, $\overline{w}_3 = [0.192, 0.25]$, $\overline{w}_4 = [0.225, 0.275]$. By an interval number ranking formula, the priority of all candidates is $w_1 = 0.296$, $w_2 = 0.346$, $w_3 = 0.11125$, $w_4 = 0.246$. From this, candidate 2 is the best, followed by candidate 1 and candidate 4, and candidate 3 is the worst.

2.4 Summary and Future Research

This chapter studies the AHP application approach based on interval number judgment matrices, which include interval number complementary judgment matrices and interval number reciprocal judgment matrices. Little research has focused on a consistency definition of the internal number reciprocal judgment matrix. This chapter put forward local consistency and consistency degree of interval number reciprocal judgment matrix on the basis of a random deterministic judgment matrix. The research on weight solution of interval number reciprocal judgment matrices is rich but ignores the consistency of the matrix. Therefore, the decision maker cannot evaluate the credibility of the results. This chapter analyzes the consistency of the matrix and builds a new weight solution model, namely, the upper and lower limit model of weight and probable value. Then, the integration approach was introduced.

As the other form of interval number judgment matrix, the research on weight solution on interval number complementary judgment matrix is scant. This chapter put forward a new weight solution approach based on the random deterministic judgment matrix. However, the consistency research on complementary judgment matrix is not fully developed so it needs further investigation. Future research is summed up in four aspects:

1. Actually, it is not suitable to define whether the judgment is consistent or not by mathematics in some scenes, because consistency is the syntheses of the decision maker's objective judgment and subjective preference. In consistency analysis,

human thought complexity and uncertainty need to be taken into account; for instance, the contradiction between inner thought and behavior, the change of decision maker utility, and so on even infiltrate into judgment matrix in a certain way. In this case, simple mathematical determination of consistency analysis is not enough. Judging from the existing research, group decision-making behavior and judgment matrix mathematics study achievements have been studied extensively and studies combining both issues would be worthwhile.

2. In weight solution models, all existing models are based on the analysis of consistency, and different consistency definitions lead to different weight analysis models. In this view, as the breakthrough on consistency definition, the weights of the method will also cause corresponding change.

3. Many methods can be used to study the consistency and weight of the interval number judgment matrix. It is vital to perform a comparative analysis of these components because it is difficult to predict or determine what policy makers think. A combination of mathematical analysis methods and experimental group decisions represents a breakthrough.

4. This chapter presents two judgment matrices. In fact, one important problem is worth pondering. What situation is suitable for any class of preference form? Do different kinds of preference really show decision-makers' judgment opinion? What kind of preference form can represent the decision maker's complex psychological behavior? Academics put forward many different views of group decision preferences. Flexibility should be further researched.

3

DECISION-MAKING METHOD OF UNASCERTAINED NUMBER COMPARISON MATRIX

3.1 Introduction

The processing and expression of unascertained information are important research topics in the information age. The uncertainty that people encounter at the earliest stage is random, and encompasses probability theory, mathematical statistics, and other subjects. Along with the development of production, science, and technology, the fuzziness resulting from unclear definition has been gradually recognized and fuzzy mathematics is the tool that handles fuzzy information. In the 1980s, scholars studied information that is partly known and partly unknown, namely, gray information. Compared with random fuzziness, gray has less known composition, the indetermination contains higher degree, and the gray set and the gray number constitute the foundation of gray mathematic theory. When engaged in construction engineering theory, Wang Guang Yuan [15] discovered an uncertainty that is different from the random, fuzziness, gray, called an unascertained property. This kind of indetermination mainly is from purely subjective understanding of indetermination that results from decision makers not knowing the complete state and mathematical relation of a situation. From actual decision-making problems, this chapter puts forward a new method of AHP. Its meaning and mathematical expression are similar to the unascertained number and then the unascertained number is adapted to AHP.

3.2 Consistency Analysis of Unascertained Number Comparison

3.2.1 Unascertained Number Comparison Matrix

In the AHP fulfillment, affected by the indeterminate factors of decision-making problems, some scholars expanded AHP to form the research field of indeterminate AHP; the most familiar indeterminate AHP includes three kinds of forms:

1. Random AHP. Considering the randomness of the decision maker's judgment, the preference number may be any one between 1/9 to 9. Estimates are expressed by the forms, $a_{ij} = a'_{ij} \pm \Delta a_{ij}, \Delta a_{ij}$, which obeys some special distribution, such as gamma distribution, lognormal distribution, etc.
2. The interval number AHP. The decision maker thinks that the estimated scope contains a higher credibility than a certain number, namely, expressing an indeterminate judgment by the interval number $\overline{a_{ij}} = [a_{ij}^L, a_{ij}^U]$;
3. Fuzzy number AHP. The decision maker thinks that a boundary expressed through a certain number is too affirmative and that it will form a misty boundary area, thus forming a fuzzy number AHP.

The AHP indetermination expanded the application of AHP but also brought some problems that need to be solved. When calculating power weight, random AHP suppose's that elements obey some continuous distribution (such as gamma distributes, lognormal distribution, etc.). This kind of distribution sometimes cannot exactly reflect decision makers' preferences. The interval number AHP and fuzzy number AHP using an interval number or fuzzy number to express judgment will increase the uncertainty of the judgment. Comparing project 1 and project 2 as an example, if the decision maker thinks the possibility is 30% that the two projects are of the same importance ($a_{12} = 1$), the possibility is 50% that project 1 is a little bit more important than project 2 ($a_{12} = 3$), but the possibility that one is more important than the other ($a_{12} = 5$) is only 20%. Hence, if we use the interval number $\overline{a_{12}} = [1,5]$ to express differences between degrees of importance, the indetermination of decision will be increased, and decision makers' preference information will not be utilized fully. A power-adding method $a_{12} = 1 \times 30\% + 3 \times 50\% + 5 \times 20\% = 2.8$ to process, so weak that the different opinions of different decision makers is not reflected, and the information on which has different possibilities to take a value inside the zone will be lost, because the error margin was larger. But if described with the binary array form, then decision makers' preference information can be further fully expressed, avoiding the shortage of existing methods. For example:

$$a_{12} = \begin{cases} (1,\, 30\%) \\ (3,\, 50\%) \\ (5,\, 20\%) \end{cases} \tag{3.1}$$

A binary array (1, 30%) means that there is a 30% possibility that the decision maker thinks that two projects contribute the same.

Along with the increment of the decision problem's complexity, group decision making has attracted the extensive attention of scholars. AHP is effective tool for processing decision-making problems, but there exist the following problems:

1. While making use of AHP to carry on group decision making, people generally adopt two kinds of different integration methods, namely, firstly integrating each decision maker's single comparison matrix to get a comprehensive comparison matrix, then solving the comprehensive comparison matrix's weight to acquire the sequence of projects or computing the weight vector of each decision-maker's comparison matrix firstly, then integrating each weight vector. Two kinds of methods both get an extensive application, but sometimes they cannot educe the same result. To a certain degree this limits the application of AHP in group decision making.

2. No matter which way the geometry average method or the weight-adding method is adopted to integrate, it cannot avoid information loss, such as the geometry average method and the weight-adding method, which are extensively used. Suppose that decision maker d gives a comparison matrix $M^d = (m_{ij}^{\,d})_{n \times n}$, $d = 1, \ldots, m$, and m is the number

of decision makers, integrating the comparison matrix of each decision maker to be $M^g = (m_{ij}^{\ g})_{n \times n}$, the result of the geometry average method is $m_{ij}^{\ g} = \left(\prod_{d=1}^{m} m_{ij}^{\ d} \right)^{1/m}$, but the result of the weight-adding method is $m_{ij}^{\ g} = \left(\sum_{d=1}^{m} m_{ij}^{\ d} \right)/m$. These integrating methods do not make good use of the information of decision makers about methods because their information consists of opinions only on alternatives. Moreover, the integrating result cannot reflect the differences between individuals; a consistent opinion is not easily reached. We can use binary arrays to indicate the judgments of the experts' team. The statistical method is used to simulate distribution of the expert's opinion, stating v^k (the frequencies $a_{ij}^{\ k}$) to a_{ij}, $k = 1, \ldots, M$, where M is the decision-making opinion number of categories, $P(a_{ij}^{\ k}) = v^k/m$, the form of binary arrays reflecting expert's opinion is

$$a_{ij} = \begin{cases} \left(a_{ij}^{\ 1}, P(a_{ij}^{\ 1}) = \dfrac{v^1}{m} \right) \\ \cdots \quad \cdots \\ \left(a_{ij}^{\ M}, P(a_{ij}^{\ M}) = \dfrac{v^M}{m} \right) \end{cases} \tag{3.2}$$

Under some decision-making scenes, the importance of two elements are objectively certain to the decision maker. But because of the condition restriction, the proof controlled by the decision maker is not true enough to ascertain the real information about alternative, and the quantitative relationships among alternatives, thus resulting in the uncertainty of judgment. For example, the decision maker cannot accurately give comparisons of the distance length between the two sides of A and B, but the distance of two sides is objectively deterministic; the purely subjective uncertainty is different from random and fuzzy. In this case, it also cannot use the existing theory method of uncertain AHP.

This chapter puts forward a new theory method of uncertain AHP that expresses the judgments of decision makers to binary arrays; the judging element of binary array form has the following characteristics:

1. The element obeys discrete distribution. Denote α for the total credibility of decision makers, if $\alpha = 1$, namely, $\sum_{k=1}^{M} P(a_{ij}^{k}) = \alpha = 1$, it shows that the estimates of judgment results by the decision makers is complete (numbers respectively are $a_{ij}^{1}, ..., a_{ij}^{M}$). The binary array form of $\alpha = 1$ belongs to a random information expression, but the international literature research [85, 87] showing that comparison matrix element obeys discrete distribution (credibility is the distribution rate) is scant.

2. If we think the credibility is $\sum_{k=1}^{M} P(a_{ij}^{k}) = \alpha < 1$, it shows that a decision maker does not confide in the result, in addition to $a_{ij}^{1}, ..., a_{ij}^{M}$, the judging credibility could also be $1 - \alpha$, expressed as:

$$a_{ij} = \begin{cases} [a_{ij}^{1}, P(a_{ij}^{1})] \\ \cdots\ \cdots \\ [a_{ij}^{M}, P(a_{ij}^{M})] \\ (\text{Other}, 1 - \alpha) \end{cases} \tag{3.3}$$

The binary array form in this case cannot be regarded as a random variable number process; AHP and its uncertain spread cannot accurately describe the uncertain judgment.

This chapter studies AHP comparison matrixes and its properties whose elements are expressed by a binary array form. Because of the concept, manifestation of the binary array form is close to unascertained number [16], and the unascertained number is introduced into AHP.

Definition 3.1

To $a = x_1 < \cdots < x_M = b$, if

$$\phi(x) = \begin{cases} x = x_i, P(x = x_i) = \alpha_i \\ \text{Other}, 0 \end{cases}, \sum_{i=1}^{M} \alpha_i = \alpha, 0 < \alpha \le 1,$$

$[a, b]$ and $\phi(x)$ constitute a "unascertained" rational number, α is the total credibility ($\alpha=1$, degenerate into random variable), $[a, b]$ is the value space, x_i is the possible value of x, α_i is the credibility of $x=x_i$, M is the rank number of the "unascertained" number.

$x_i = [x_i^L, x_i^U]$ is used to denote x_i, x_i^L is the bottom limit, x_i^U is the upper limit, then $[a, b]$ and $\phi(x)$ constitute a blind number.

The definition and property of the unascertained number comparison matrix is given below.

Definition 3.2

A matrix made up of unascertained rational numbers is called an unascertained rational number comparison matrix, marked as $\hat{A} = (\widehat{a_{ij}})_{m \times n}$,

$$\widehat{a_{ij}} = \begin{cases} [a_{ij}^1, P(a_{ij}^1)] \\ \cdots \\ [a_{ij}^M, P(a_{ij}^M)] \end{cases},$$

a_{ij}^k is one of the M possible values of $\widehat{a_{ij}}, k = 1,\ldots, M$ and $P(a_{ij}^k)$ represents the assurance of a_{ij}^k.

Property 3.1

① $0 < P(a_{ij}^k) \leq 1$, the total credibility $\alpha = P(a_{ij}^1) + \cdots + P(a_{ij}^M)$ satisfies $0 < \alpha \leq 1$;

② According to AHP 1–9 scale, $a_{ij}^k > 0, a_{ij}^k \in \{1/9, \ldots, 9\}$, $\widehat{a_{ii}} = 1, P(\widehat{a_{ii}} = 1) = 100\%$;

③ Reciprocality, $\widehat{a_{ji}} = 1/\widehat{a_{ij}}$,

$$\hat{a}_{ji} = \begin{cases} \left(\dfrac{1}{a_{ij}^1}, P(a_{ij}^1) \right) \\ \cdots \\ \left(\dfrac{1}{a_{ij}^M}, P(a_{ij}^M) \right) \end{cases}.$$

In Definition 3.2, if $[a_{ij}^{kL}, a_{ij}^{kU}]$ is used to denote a_{ij}^{k}, constituting an AHP comparison matrix with an ascertained number. When $a_{ij}^{1} = \cdots = a_{ij}^{M}$, $P(a_{ij}) = 100\%$, abbreviated to the form of $(a_{ij}, 100\%)$, then the unascertained number comparison matrix deteriorates into a determinate comparison matrix.

If no confusion is created, the unascertained rational number comparison matrix and blind number AHP comparison matrix are uniformly called the unascertained number analytical hierarchy process (UNAHP).

During group decision making, the credibility of an unascertained number comparison matrix can adopt to the method of frequency stating. For a single decision maker, the cheng (one tenth) concept can denote the confidence of the decision maker. The minimum confidence is defined as one cheng or tenth, the biggest confidence is defined as ten cheng, the corresponding credibility number is between 0.1 and 1, and the value is separated into 0.1, 0.2, 0.3, 0.4, 0.5, 0.6, 0.7, 0.8, 0.9, 1. For example, when two elements compare, if there is a 30% possibility that the decision maker thinks that element 1 and element 2 are equally important, then $P(a_{12}=1)=0.3$.

When applying UNAHP to make a decision, the structure establishment method is the same as AHP. The consistency-checking methods and weight-calculating methods of the unascertained number comparison matrix are discussed as follows.

3.2.2 Consistency Definition of Unascertained Number Comparison Matrix

In the unascertained number comparison matrix, there is a deterministic comparison matrix $A = (a_{ij})_{n \times n}$, $a_{ij} \in \hat{a}_i$ that can reflect the true preference of the decision maker. In practice, a deterministic comparison matrix can be randomly generated in an unascertained number comparison matrix to approximately estimate a decision maker's preference. Thereafter, a random deterministic comparison matrix concept of unascertained number comparison matrix is put forward, based on the consistency characteristics of unascertained number comparison matrix $\hat{A}$.

Definition 3.3

$A=(a_{ij})_{n\times n}$ is called a random deterministic comparison matrix of unascertained number comparison matrix $\hat{A}$. A is constructed as follows: when $1 \leq i \leq j \leq n$, $a_{ij}= a_{ij}^k$ (if $\widehat{a_{ij}}$ is the blind number, then $a_{ij} \in a_{ij}^k = [a_{ij}^L, a_{ij}^U]^k$, a_{ij} is generated randomly in $[a_{ij}^L, a_{ij}^U]^k$ according to the probability of uniform distribution). The value of k is determined by $P(a_{ij}^k)$ with the method of roulette, and if $\alpha = 1$, $k \in \{1, \dots , M\}$ if $\alpha < 1$, $k \in \{1, \dots , M + 1\}$. Make a_{ij}^{M+1} to take any value in $\{1/9, \dots , 9\}$, $P(a_{ij}^{M+1}) = 1 - \alpha. \alpha_{ji} = 1/ a_{ij}, a_{ii} = 1, i, j = 1, \dots, n$.

Definition 3.4

For the unascertained number comparison matrix $\hat{A}$, a deterministic random comparison matrix is generated by Definition 3.3. If some deterministic comparison matrix uncertainty has a consistent property, then $\hat{A}$ has local consistency.

A judging element a_{ij} is generated by Definition 3.3; it can reflect the decision makers' judgments based on some credibility; the random deterministic comparison matrix A, which is composed of a_{ij}, can reflect the local characteristics of $\hat{A}$. Therefore, $\hat{A}$ has the characteristics of local consistency if there is some random matrix A having the consistency.

As the complexity of decision-making problems increases and information is incomplete, and $\hat{A}$ cannot achieve local consistency, the definition of local satisfactory consistency of the unascertained number comparison matrix is given.

Definition 3.5

A deterministic random comparison matrix is generated by Definition 3.3. If matrix A has satisfactory consistency $(CR(A) \leq 0.1)$, then $\hat{A}$ is said to have local satisfactory consistency. Otherwise, $\hat{A}$ does not have local satisfactory consistency.

If $\hat{A}$ has local consistency, it must have a local satisfactory consistency. Conversely, if $\hat{A}$ does not have a local satisfactory consistency, the weight that is exported by $\hat{A}$ is unreliable, and the results cannot serve for decision making.

Any deterministic matrix randomly generated by Definition 3.3 can locally reflect characteristics of $\hat{A}$. In practice, a consistency matrix randomly generated can be the approximate decision maker preferences. For the randomly generated N (N is sufficiently large) deterministic matrix, if the larger the number of comparison matrices t satisfies $(CR(A) \leq 0.1)$, the more the matrices can be the approximate decision maker preferences; this indicates that the consistency of $\hat{A}$ is better, otherwise the worse. Therefore, the consistency degree of $\hat{A}$ can be reflected by $\eta = t/N \times 100\%$.

Definition 3.6

N deterministic matrix A is randomly generated by Definition 3.3, denoted by $A^1, \ldots, A^N$. If t deterministic matrices A have satisfactory consistency $(CR(A) \leq 0.1)$, then the consistency degree η of $\hat{A}$ can be denoted as

$$\frac{t}{N} \times 100\%.$$

If $\eta = 0$, $\hat{A}$ has a local satisfactory consistency; theoretically, as long as $\eta > 0$, the matrix contains the true preferences of decision makers in the comparison matrix. When the consistency degree η is small (generally $\eta \geq 60\%$), the consistency level of $\hat{A}$ is poor; otherwise, good ($\eta \geq 60\%$). If η is larger, the opinions of experts is concentrated more. When the consistency degree of the comparison matrix is poor, policy makers need to reevaluate or make the appropriate improvements.

3.3 Weight Model of Unascertained Number Comparison Matrix

3.3.1 Operation Rule of Unascertained Number

1. Let the unascertained number

$$\hat{Q}_A = \begin{cases} [x_1, f(x_1)] \\ \cdots \\ [x_n, f(x_n)] \end{cases}, \quad \hat{Q}_B = \begin{cases} [y_1, g(y_1)] \\ \cdots \\ [y_m, g(y_m)] \end{cases},$$

The matrix shown in Figure 3.1 shows the possible values for the band edge sum matrix; the matrix shown in Figure 3.2 shows

x_1	$x_1 + y_1$	$x_1 + y_2$	$\cdots$	$x_1 + y_m$
$\cdots$	$\vdots$	$\vdots$	$\vdots$	$\vdots$
x_n	$x_n + y_1$	$x_n + y_2$	$\cdots$	$x_n + y_m$
$+$	y_1	y_2	$\cdots$	y_m

Figure 3.1 Possible value matrix with side addition.

$f(x_1)$	$f(x_1)g(y_1)$	$f(x_1)g(y_2)$	$\cdots$	$f(x_1)g(y_m)$
$\cdots$	$\vdots$	$\vdots$	$\vdots$	$\vdots$
$f(x_n)$	$f(x_n)g(y_1)$	$f(x_1)g(y_2)$	$\cdots$	$f(x_1)g(y_m)$
x	$g(y_1)$	$g(y_2)$	$\cdots$	$g(y_m)$

Figure 3.2 Reliability matrix with side multiplication.

the credibility for the band edge product matrix [16]. Rank the $n \times m$ possible values $x_i + y_j$ in Figure 3.1 in order from small to large (the same elements combined), denoted by $v_1, \ldots , v_l$. Arrange the corresponding $n \times m$ credibility $f(x_i)$ $g(y_j)$ in Figure 3.2 in a sequence (the same credibility of the element may add up), denoted by $h(v_1), \ldots , h(vl)$, then

$$Q_A + Q_B = \begin{cases} [v_1, h(v_l)] \\ \cdots \\ [v_l, h(v_l)] \end{cases} , \quad l \leq n \times m.$$

When computing the high-rank unascertained number, the method of combined reliabilities can be used to reduce operations. The subtraction, multiplication, and division operations of unascertained number change the $+$ in Figure 3.1 into $-$, $\times$, $\div$, the credibility of the band edge product matrix in Figure 3.2 does not change.

2. High unascertained rational number reduction method: combined low reliability points [16] for unascertained rational number $\{[x_1, x_n], f(x)\}$, make $x_1 < x_2 < \ldots < x_k \leq \bar{x} < x_{k+1} < \ldots < x_n$, so $\bar{x} = x_1 + x_n/2$. If $f(x_i) \leq r$ (such as access $r \leq 0.005$), the reliability of the point x_i can be thought small. Adding credibility to the left or right of the point, the method is as follows: when $2 \leq i \leq k-1$, make $f(x_{i+1}) = f(x_{i+1}) + f(x_i)$, at the same time give up the

point x_i; when $i = k$, on $i = k + 1$, make $f(\bar{x}) = f(x_k) + f(x_{k+1})$, at the same time discard point x_k or x_{k+1}; when $k + 1 < i \leq n - 1$, make $f(x_{i-1}) = f(x_{i-1}) + f(x_i)$, at the same time give up points x_i. This method does not give up the two end points x_1, x_n; the value interval of the unascertained number has not been changed.

The weight calculation method of UNAHP comparison matrix is analyzed below. The methods based on unascertained number algorithm and Monte Carlo method are proposed. Two methods take full advantage of decision makers' information, which can get more detailed results than the interval methods.

3.3.2 Weight Model Based on Operation Rule of Unascertained Number

The ideas of solving the weight based on operation rule of unascertained number are summarized as follows: According to the operation rule of unascertained number, for UNAHP comparison matrix $\hat{A}$, make its elements directly involved in the ranking calculation formula of AHP, and thereby the weight of the unascertained number comparison matrix is derived. Take, for example, the AHP line and normalized ranking method. Its weight calculation formula is $w_i = \sum_{j=1}^{n} a_{ij} / \sum_{i=1}^{n} \sum_{j=1}^{n} a_{ij}$, for $\hat{A}$, take $\widehat{a_{ij}}$ into the above formula, obtain $\widehat{w_i} = \sum_{j=1}^{n} \widehat{a_{ij}} / \sum_{i=1}^{n} \sum_{j=1}^{n} \widehat{a_{ij}}$, $i = 1, \dots, n$, $\widehat{w_i}$ is the unascertained number, seek the mathematical expectation $E(\widehat{w_i})$, the expectable weight of the project can be obtained.

3.3.3 Weight Model Based on Monte Carlo

The Monte Carlo method has been widely used in dealing with uncertain decision-making problems. In this chapter, a method to solve the weight of UNAHP comparison matrix is designed based on existing research.

For the unascertained number comparison matrix $\hat{A}$, N random deterministic matrixes A^I are generated by Definition 3.3, $I = 1, \dots, N$, if $CR(A^I) \leq 0.1$, the corresponding weights $w^I = (w_1^I, \dots, w_n^I)^T$ are solved, denoted as $w_i^L = \min_{I=1,\cdots,N} w_i^I$, $w_i^U = \max_{I=1,\cdots,N} w_i^I$, the weight can be expressed by interval number $\overline{w}_i = [w_i^L, w_i^U]; w_i^I$ is denoted as the weight of the ith project in the Ith comparison matrix, $Rank_i^j$ is the cumulative number of w_i^I ranking number j and t is the number of matrices that have a satisfying consistency ($CR(A^I) \leq 0.1$); the mean values of $Rank_i^j$, $\eta = t/N$, and w_i are calculated.

Monte Carlo simulates decision makers' different values based on the size of credibility, and then the random deterministic comparison matrix is created. The arrangement of the program's weight and its average are counted by the weight of the random deterministic comparison matrix. This method requires a larger number of generated matrices, a large amount of calculation, and the results with interval numbers.

For the matrix containing part of the unascertained number and part of the determined number, the methods based on the unascertained number method and Monte Carlo algorithm method are also applicable.

3.4 Weight Integration Method of Unascertained Number Comparison

3.4.1 Weight integration Based on Operation Rule of Unascertained Number

For the random deterministic comparison matrix obtained by each criterion, the weight form is also an unascertained number. Therefore, the algorithms of the unascertained number can be used to calculate the weight combination.

3.4.2 Weight Integration Based on Monte Carlo

In the given hierarchical structure by decision makers, m unascertained number comparison matrixes is obtained by the pairwise comparison of all of the criteria and subcriteria, denoted by $\hat{A}_k$, $k = 1, \ldots, m$, so the weight combination is calculated as follows:

Step 1: For $\hat{A}_k$, $k = 1, \ldots, m$, N random deterministic comparison matrices are generated by Definition 3.3, comparison matrix A_{ky}, which satisfies $CR(A^I) \leq 0.1$, and the numbers recorded, $y = 1, \ldots, t_k$. The consistency degree is calculated. $\eta = \min\limits_{k=1,\cdots,m} (\eta_k)$ is used to measure the total consistency degree of the unascertained number comparison matrix.

Step 2: For $k = 1, \ldots, m$, y is generated randomly in $\{1, \ldots, t_k\}$, the weight of comparison matrix A_{ky} is solved. Then, according to the hierarchical structure of A_{ky}, the weighting combination formula is used to calculate the combination weight combined weights $w = (w_1, \ldots, w_n)^T$ are obtained.

Step 3: Repeat step 2 N times, and the weights of the deterministic comparison matrix are obtained, denoted as $w_i^I, I = 1, \ldots, N$. Set $w_i^L = \min_{I=1,\cdots,N} w_i^I, w_i^U = \max_{I=1,\cdots,N} w_i^I$, then the weight can be expressed by the interval number $\overline{w}_i = [w_i^L, w_i^U]$. Record $Rank_i^j$ as the cumulative number of the weight w_i^I for project i, which is ranking number j; $Rank_i^j$ and the average of w_i are calculated. Decision makers can use $\overline{w}_i = [w_i^L, w_i^U]$, $Rank_i^j$, and the average of w_i to rank the priority order of these projects.

3.5 Example Analysis

Example 3.1

Select the most appropriate director of the office from three candidates. The evaluation criteria are ability to work, level of education, and interpersonal relationships. Assuming a comparison based on the criterion of ability to work, an unascertained rational number matrix $\hat{A}$ is obtained (lower triangular part of the elements is given from reciprocity; $\widehat{a_{ij}}$ is three-order unascertained rational numbers).

$\widehat{a_{12}}$	$\widehat{a_{13}}$	$\widehat{a_{23}}$
(4, 70%)	(1, 60%)	(1/8, 50%)
(6, 10%)	(1/2, 30%)	(1/6, 20%)
(1, 20%)	(2, 10%)	(1/5, 30%)

Suppose that N=200, the coincidence degree $\eta = 73.5\%$ of $\hat{A}$ is obtained, the value is larger, indicating their consistency degrees are better. Results-based unascertained number algorithms are shown in Table 3.1. The possibility of candidate 1's weight taking 0.4394 is 56%; the possibility of candidates 2 and 3 weights taking 0.4778, 0.5974 is 40%; candidate 1's and 3's weights of mathematical expectation are closer; candidate 3 is slightly better than candidate 1; candidate 2's weight is obviously less than those of 1 and 3; the priority of the three candidates is $w_3 > w_1 > w_2$.

Table 3.1   Weight Based on Unascertained Number Operation Rule

WEIGHT RANK	$\widehat{W}_1$	$\widehat{W}_2$	$\widehat{W}_3$
1	(0.271, 26%)	(0.0808, 56%)	(0.353, 20%)
2	(0.4394, 56%)	(0.1175, 30%)	(0.4478, 40%)
3	(0.6154, 18%)	(0.1534, 14%)	(0.5974, 40%)
$E(w_i)$	0.4273	0.102	0.5007

Table 3.2   Weight Based on Monte Carlo Simulation

	$\overline{W}$	MEAN	$RANK_i^1$	$RANK_i^2$	$RANK_i^3$
$\overline{W}_1$	[0.2247, 0.6316]	0.4210	38	109	0
$\overline{W}_2$	[0.0526, 0.1655]	0.0858	0	0	147
$\overline{W}_3$	[0.3158, 0.6098]	0.4932	109	38	0

Simulation results from using the Monte Carlo method 200 times are shown in Table 3.2. From Table 3.2, in work capacity criteria, based on average weight, the priority order of the three candidates is $w_3 > w_1 > w_2$; candidates 1 and 3 are close in weight; candidate 1 is ranked number 1 is 38 times; candidate 3 is ranked number 1 109 times; and the weight of the candidate 2 is obviously less than candidates 1 and 3.

The priority of candidates based on work capacity criterion is the same by the Monte Carlo method and a method based on the operation rule of unascertained numbers. In this case, if using the weighted method to process, then one obtains an interval comparison matrix (upper triangle elements are $a_{12} = 3.6$, $a_{13} = 0.95$, $a_{23} = 0.156$); the weights are $w_1 = 0.402$, $w_2 = 0.0937$ $w_3 = 0.5043$. The average results are similar to results determined by the method discussed in this chapter. If using the interval number method, an interval number comparison matrix is obtained (upper triangle elements are $\overline{a}_{12} = [1, 6]$, $\overline{a}_{13} = [1/2, 2]$, $\overline{a}_{23} = [1/8, 1/5]$). Its weight is $w_1 = [0.2305, 0.5594]$, $w_2 = [0.068, 017]$, $w_3 = [0.3589, 0.623]$ unascertained number comparison matrix method in this chapter can provide decision makers with more extensive decision-making information.

Example 3.2

The issue is the same as in Example 3.1 with four candidates. The distributions of 30 experts' preferences are counted based on work capacity criteria (suppose the experts' credibility is $\alpha = 1$), and the blind number AHP matrix is obtained.

$\widehat{a}_{12}$	$\widehat{a}_{13}$	$\widehat{a}_{14}$	$\widehat{a}_{23}$	$\widehat{a}_{24}$	$\widehat{a}_{34}$
([3-4], 50%)	([1-2], 30%)	([2-3], 20%)	([1/6-1/5], 60%)	([1/2-1], 70%)	([3-4], 50%)
([2-3], 30%)	([1/2-1], 30%)	([3-4], 60%)	([1/2-1, 20%)	([2-3], 10%)	([2-3], 40%)
([1/2-1], 20%)	([2-3], 40%	([1/2-2], 20%)	([1/4-1/3], 20%)	([1-2], 20%)	([1-2], 10%)

When $N = 200$, the consistency degree of $\hat{A}$ is $\eta = 70\%$; this consistency value is good. Simulation results using the Monte Carlo method 200 times are shown in Table 3.3. From Table 3.3, for the work capacity criterion, based on average weight, the priority of the candidates is $w_1 > w_3 > w_4 > w_2$; candidate 1 is obviously better than the other three candidates; candidates 2 and 4 are close in weight; in 200 times simulation, candidate 2 with 51 times outnumbers candidate 4; candidate 4 with 89 times outnumbers candidate 2, indicating that candidate 4 is slightly better than candidate 2.

Decision makers can use the same method to obtain candidates' weights for the criteria of level of education and interpersonal relationships. Then according to the AHP theory, the combined weight of some candidates is obtained to provide a reference for decision making.

The introduction of unascertained number into the AHP is a useful supplement in the AHP study, which could meet some needs of decision making under uncertainty situations. Its consistency test and the weight calculation are simple and practical. For this new method of AHP, other weight calculation methods and other effective ways to improve comparison matrices with poor consistency need further research.

Table 3.3   Weight Based on Monte Carlo Simulation

	$\overline{W}$	MEAN	$RANK_i^1$	$RANK_i^2$	$RANK_i^3$	$RANK_i^4$
$\overline{W}_1$	[0.2350, 0.5223]	0.4053	90	50	0	0
$\overline{W}_2$	[0.0800, 0.2121]	0.1164	0	0	51	89
$\overline{W}_3$	[0.1869, 0.5005]	0.3525	50	90	0	0
$\overline{W}_4$	[0.0809, 0.2135]	0.1257	0	0	89	51

3.6 Summary and Future Research

This chapter is against the notion that the existing AHP method in some decision-making circumstances cannot accurately reflect the preferences of decision makers and the information loss problem arising from experts' preference aggregation in group decision making. A new type of uncertainty AHP of which the elements obey the discrete distribution is researched, and the concept of unascertained number is introduced to the AHP. The ideology of local consistency and consistency degree is used to test the unascertained number comparison matrix. According to the properties of unascertained number, two weight calculation methods are put forward based on unascertained number of algorithms and Monte Carlo. Finally, the feasibility of the application of this method is shown through two examples. From the point of information expression, the unascertained number can objectively express preferences of decision makers. From the perspective of group decision making, it is also a better form of expression of group preferences. Further research involves two aspects:

1. Simple consistency measurement and weight-ranking methods should be studied. The method proposed in this chapter has a huge workload, so a more convenient way needs to be studied, and development calculation software based on preferences of unascertained number is imminent.
2. The unascertained number can contain a wealth of information, but in group decision making, different decision makers may have diverse information. How to aggregate such information, which has a large difference between preferences, such as the transformation and aggregation of unascertained numbers, interval numbers, and fuzzy numbers, and other issues, need to be considered.

4

DECISION-MAKING METHOD OF THREE-POINT INTERVAL NUMBER COMPARISON MATRIX

4.1 Introduction

Because of the difference of knowledge structure, judgment level, and individual preference, decision makers would probably express their judgment preference via the different structures in the process of group decision making such as utility, comparison, and other factors. Due to the complexity and uncertainty of the decision-making problem and the fuzziness of human beings' thought, it is unrealistic to depict complex problems via the certain preference style. Some scholars focus on the uncertainty decision-making problem, and some uncertain mathematic methods such as interval number, fuzzy number, random theory, and unascertained number are widely used in the decision-making field. Based on existing studies, the three-point interval number comparison matrix is suggested to express the decision maker's uncertain preference.

According to the decision-making preference structure, the group decision-making aggregation approach includes the aggregation on the same kind of preference structure and the different kind of preference structure. The study on the aggregation of the same kind of preference structure has received much attention, but study of the aggregation of the different kinds of preference structures is still a new field. Decision makers would probably provide different preference structures due to the difference of their knowledge structure, judgment level, and individual preference. In addition, due to the development of intelligent decision-making support systems integrated by communications technology and computer technology, the practicality and feasibility of group decision-making technology should be improved. A new way to express the decision maker's preference via the three-point interval number comparison matrix is put forward. In addition, the approach on the aggregation of uncertain preference based on the three-point interval number reciprocal comparison matrix and complementary comparison matrix is studied.

This chapter is structured as follows. The new preference style of the three-point interval number comparison matrix is put forward in the second section. The consistency of the three-point interval number reciprocal comparison matrix and the weight estimation approach are developed. In the third section, the three-point interval number complementary comparison matrix is put forward, and an approach of how to aggregate two kinds of three-point interval number comparison matrix is suggested. The application steps via the example are explained in the fourth section.

4.2 Application of Three-Point Interval Number Reciprocal Comparison Matrix

4.2.1 Definition of Three-Point Interval Number Reciprocal Comparison Matrix

Due to the uncertainty of decision-making problems, scholars expanded the analytical hierarchy process (AHP) by techniques such as interval AHP, fuzzy AHP, and random AHP. The AHP application scope is extended via the strength of these approaches. However, the deficiency of these approaches is that the comparison elements should obey a certain continuous distribution such as estimating the weight vector. The continuous distribution assumption cannot reflect the decision makers' preference in particular case. The comparison between alternative 1 and alternative 2 is as an example. Suppose that the decision makers have a preference as follows. The possibility that two alternatives hold similar importance is 10% (that is, $a_{12} = 1$), the possibility that alternative 1 is a little more important than alternative 2 is 80% (that is, $a_{12} = 3$), and the possibility that alternative 1 is obviously more important than alternative 2 is 10% (that is, $a_{12} = 5$). In this situation, using the weighing method as $a_{12} = 1 \times 10\% + 3 \times 80\% + 5 \times 10\% = 3$ to express the decision maker's preference can counterbalance the random error to some extent, but the counterbalance reliability using 3 to express a_{12} is not completely certain to the decision maker. If the interval number $\overline{a_{12}} = [1, 5]$ is used to express the different importance of the two alternatives, it would increase the uncertainty of decision making, and it does not use the decision maker's judgment information completely. The three-point interval number is suggested to express the decision makers' uncertain preference. The new interval number style $\overline{a_{12}} = [1, 3, 5]$ is adopted. The value 3 is the most possible judgment and the judgment possibility is 80%.

In addition, the group decision-making process is focused on by scholars as the complexity of decision-making problems. AHP is a useful tool to deal with group decision-making problems, but some problems still exist. For instance, the information is lost while aggregating decision makers' preference and this limits the AHP's application. This chapter proposes using the three-point interval number to express the group's preference to handle the information lost problem. Assume m as the number of decision makers, and the distribution of expert preference can be fitted via the statistical method. For judgment c_{ij}, the frequency of c_{ij}^k is v^k, $k = 1, \ldots, M$, where the value M represents the classification numbers of decision-making judgment preference. The formula $P(c_{ij}^k) = v^k / m$ is holding, let $c_{ij}^L = \min_k c_{ij}^k$, $c_{ij}^U = \max_k c_{ij}^k$, $c_{ij}^M = c_{ij}^f$, and let $P(c_{ij}^f) = \max_k \{P(c_{ij}^k)\}$. As a result, the decision-making group's preference can be expressed as $[c_{ij}^L, c_{ij}^M, c_{ij}^U]$.

Definition 4.1

The interval number $[c_{ij}^L, c_{ij}^M, c_{ij}^U]$ is called the three-point interval number if the relationship of $c_{ij}^L \leq c_{ij}^M \leq c_{ij}^U$ is satisfying. The value c_{ij}^L is the lower possible value of the judgment, the value c_{ij}^U is the upper possible value, and c_{ij}^M is the most possible judgment value.

How does the decision maker apply the three-point interval number in process decision making? For a single decision maker, the decision maker can adopt the two-point interval number to estimate the upper and lower value $[c_{ij}^L, c_{ij}^U]$. Meanwhile, he or she should provide the most possible judgment c_{ij}^M according to his estimation (Figure 4.1).

In the background of group decision making, using the statistics approach to fit decision makers' preferences and the three-point interval number judgment can be obtained (see Figure 4.2). The distribution probability of the most possible value $P(c_{ij}) \geq \delta, \delta$ is a constant, and its value should be decided by the decision makers' experience. Only when δ reaches a certain degree can it be called the most possible

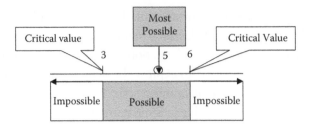

Figure 4.1 Application method of three-point interval number.

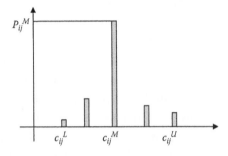

Figure 4.2 Application method of three-point interval number for group.

value. In group decision making, if $\delta < 60\%$, this indicates that the group preference is too disperse, and decision makers should examine the decision-making problem again and make some adjustments. If the precision of decision making is not high, the most possible value can also be calculated as

$$c_{ij}^{M} = \frac{c_{ij} + \cdots + c_{ij}^{m}}{m}.$$

Using the two-point interval number, the judgment interval may be too large to cover the whole scope of the possible judgment and would increase the decision-making uncertainty. Compared with the traditional two-point interval number, the three-point interval number not only keeps the interval scope but also makes the gravity spot (most possible value of a decision maker's judgment) of the most possible value prominent. It can express the decision makers' preference more accurately and make up the deficiency of the two-point interval number to some extent. The three-point interval number is similar to the triangle fuzzy number, but the latter assumes a linear progressive change between c_{ij}^{L} and c_{ij}^{M} (or c_{ij}^{U} and c_{ij}^{M}). This assumption lacks the theory basis in some decision-making processes. In addition, its assumption may be too strict to express the decision makers' real preference. In the three-point interval number $[c_{ij}^L, c_{ij}^M, c_{ij}^U]$, the judgment possibility of c_{ij}^M is highest, whereas the possibility of other values is unknown.

Definition 4.2

The comparison matrix $\overline{C} = (c_{ij}^L, c_{ij}^M, c_{ij}^U)_{n \times n}$ is called the three-point interval number reciprocal comparison matrix, where $c_{ij}^L \leq c_{ij}^M \leq c_{ij}^U$, $\overline{c}_{ji} = [1/c_{ij}^U, 1/c_{ij}^M, 1/c_{ij}^L]$, $\overline{c}_{ii} = [1,1,1]$. It is a new uncertain preference through the three-point interval number composed of the upper value, lower value, and most possible value. The decision makers' real preference is among the upper value, lower value, and most possible value, but it is difficult to determine. One cannot simply split the three-point interval number comparison matrix into a matrix composed of the lower judgment value, most possible value, and upper value. In addition, the comparison matrix composed of upper and lower judgment may be not consistent.

The application steps of three-point interval number reciprocal comparison matrix are the same with traditional AHP. First, establish the appraisal index system of the decision-making problem. Then, compare two alternatives to form the three-point interval number comparison matrix. Next, analyze its consistency and estimate the weight. According to the representation and definition of three-point interval number reciprocal comparison matrix, it is unable to measure its consistency and estimate the weight via the traditional method.

4.2.2 Consistency and Weight Estimation of Three-Point Interval Number Comparison Matrix

Definition 4.3

The set of $w_i, i = 1, \dots, n$ is denoted as the weight of three-point interval number reciprocal comparison matrix. If Formula (4.1) is satisfied, the three-point interval number reciprocal comparison matrix is completely consistent. If the $w_i, i = 1, \dots, n$ does not satisfy Formula (4.1), the three-point interval number reciprocal comparison matrix is not completely consistent.

$$c_{ij}^L \leq c_{ij}^M = \frac{w_i}{w_j} \leq c_{ij}^U, \; i, j = 1, \dots, n, \; i \neq j \qquad (4.1)$$

If matrix $\overline{C}$ is completely consistent, the formula

$$c_{ij}^M = \frac{w_i}{w_j}$$

is satisfied, which means that the decision makers' judgment is correspondent with the most possible value. If $\overline{C}$ not completely consistent,

$$c_{ij}^M \neq \frac{w_i}{w_j}, \quad \text{or} \quad c_{ij}^L \geq \frac{w_i}{w_j}, \quad \text{or} \quad c_{ij}^U \leq \frac{w_i}{w_j}.$$

To the relationship of

$$c_{ij}^M \neq \frac{w_i}{w_j},$$

the error variables cpo_{ij}, cdo_{ij} are introduced, satisfying $cpo_{ij} \geq 0$, $cdo_{ij} \geq 0$, $cpo_{ij} \times cdo_{ij} = 0$. Then, Formula (4.2) is satisfied.

$$c_{ij}^M w_j - w_i + cpo_{ij} - cdo_{ij} = 0, \, i, j = 1, \ldots, n, \, i \neq j \qquad (4.2)$$

If Formula (4.2) does not satisfy

$$c_{ij}^L \leq \frac{w_i}{w_j} \leq c_{ij}^U,$$

the error variables cp_{ij}, cd_{ij} are introduced, where $cp_{ij} \geq 0$, $cd_{ij} \geq 0$. Then, Formula (4.3) is satisfied.

$$\begin{cases} c_{ij}^L w_j \leq w_i + cp_{ij}, \, i, \, j = 1, \ldots, n, \, i \neq j \\ w_i \leq c_{ij}^U w_j + cd_{ij}, \, i, \, j = 1, \ldots, n, \, i \neq j \end{cases} \qquad (4.3)$$

The lower the value of $\sum_{i,j} cpo_{ij} + cdo_{ij}$ is, the less the deviation distance from the most possible value c_{ij}^M is. In addition, the lower the value of $\sum_{i,j} cp_{ij} + cd_{ij}$ is, the less the deviation distance from the upper and lower $[c_{ij}^L, c_{ij}^U]$ provided by decision makers is. The better the consistency of three-point interval number comparison matrix is, the lower the value of $\sum_{i,j} cpo_{ij} + cdo_{ij}$ and $\sum_{i,j} cp_{ij} + cd_{ij}$. One can record the error as $\sum_{i,j} s(cpo_{ij} + cdo_{ij}) + t(cp_{ij} + cd_{ij})$; s, t are constants to show the priority of these two errors. If $\sum\limits_{i,j} s(cpo_{ij} + cdo_{ij}) + t(cp_{ij} + cd_{ij}) \to \min$, the consistency of the three-point interval number comparison matrix is better. Based on these ideas, $P_{4.1}$ is suggested to estimate the weight of the three-point interval number comparison matrix.

$$\min c = \sum\limits_{i,j} s(cpo_{ij} + cdo_{ij}) + t(cp_{ij} + cd_{ij}) \tag{4.4}$$

$$c_{ij}^M w_j - w_i + cpo_{ij} - cdo_{ij} = 0, \; i, j = 1, \ldots, n, \; i \neq j \tag{4.5}$$

$$c_{ij}^L w_j \leq w_i + cp_{ij}, \; i, j = 1, \ldots, n, \; i \neq j \tag{4.6}$$

$$w_i \leq c_{ij}^U w_j + cd_{ij}, \; i, j = 1, \ldots, n, \; i \neq j \tag{4.7}$$

$$\sum\limits_{i=1}^{n} w_i = 1 \tag{4.8}$$

$$w_i \geq 0, \; cpo_{ij} cdo_{ij} = 0, \; cp_{ij}, \; cd_{ij}, \; cpo_{ij}, \; cdo_{ij} \geq 0, \; i, j = 1, \ldots, n \tag{4.9}$$

In $P_{4.1}$, Formula (4.4) indicates seeking a set of weights in which the error sum deviating from the upper–lower and possible value is smallest. Formula (4.5) indicates the deviation relationship of the weight and possible value. Formulas (4.6)–(4.7) indicate the deviation relations of weight and upper–lower value. Formula (4.8) indicates that the weight satisfies the normalization condition. Formula (4.9) indicates that the weight and error variables are not negative.

The optimal value of $P_{4.1}$ can be recorded as c^*, and then the following conclusions can be obtained.

Theorem 4.1
If $c^* = 0$, the three-point interval number comparison matrix $\overline{C}$ is completely consistent. Otherwise, it is also satisfied.

Proof
If $c^* = 0$, then the relation of $\sum\limits_{i,j} s(cpo_{ij} + cdo_{ij}) + t(cp_{ij} + cd_{ij}) = 0$ is satisfied. Then, $cp_{ij} = 0, cd_{ij} = 0, cpo_{ij} = 0, cdo_{ij} = 0$ can be obtained owing to

the condition of $cp_{ij}, cd_{ij} cpo_{ij}, cdo_{ij} \geq 0$. So Formula (4.1) must be satisfied. If C is completely consistent, according to Definition 4.3, Formula (4.1) must be satisfied. It is easy to obtain the conclusion of $c^* = 0$.

Theorem 4.2

If $c^* > 0$, the three-point interval number comparison matrix $\overline{C}$ is not completely consistent. The greater value the of c^* is, the less consistency of the comparison matrix $\overline{C}$ is.

Proof

If $c^* > 0$, $\forall i,j$ the value of cp_{ij}, cd_{ij}, cpo_{ij}, cdo_{ij} must not be a zero value according to Theorem 4.1. Then, matrix C is not completely consistent. Due to the formula $c^* = \sum_{i,j} s(cpo_{ij} + cdo_{ij}) + t(cp_{ij} + cd_{ij})$, the greater c^* is, the larger the degree of deviation from Formula (4.1) is, and the less the consistency of the three-point interval number comparison matrix is.

According to Theorem 4.2, c^* can be regarded as the consistency index of the three-point interval number comparison matrix.

Theorem 4.3

There must be the optimal value from model $P_{4.1}$.

Proof

It is easy to prove that the feasible domain of $P_{4.1}$ must be nonempty. One can estimate the weight that satisfies the restraint condition of $P_{4.1}$. Therefore, there must be an optimal value.

Theorem 4.4

Among the optimal value in $P_{4.1}$, if $\sum_{i,j} cpo_{ij} + cdo_{ij} = 0$, the three-point interval number comparison matrix must be consistent completely. Meanwhile, the relationship of $\sum_{i,j} cp_{ij} + cd_{ij} = 0$ must be satisfied.

Proof

It is easy to prove from formulas (4.1)–(4.3).

One can check the result from model $P_{4.1}$ based on Theorem 4.4. For convenience, generally, one can let $s = t = 1$. Formulas (4.6)–(4.7) are not a redundancy restraint condition compared with Formula (4.5) since the result holding the smallest deviation distance of the possible value may be outside of $[c_{ij}^L, c_{ij}^U]$.

Example 4.1

The three-point interval number comparison matrix is

$$\overline{C} = \begin{bmatrix} [1,\,1,\,1] & [2,\,3,\,4] & [1,\,2,\,4] \\ [1/4,\,1/3,\,1/2] & [1,\,1,\,1] & [2,\,3,\,4] \\ [1/4,\,1/2,\,1] & [1/4,\,1/3,\,1/2] & [1,\,1,\,1] \end{bmatrix}.$$

If one estimates $P_{4.1}$ without taking the restraint condition of Formulas (4.6)–(4.7) into consideration, the results of $w_1 = 0.692$, $w_2 = 0.231$, $w_3 = 0.077$ can be obtained. One can make a reverse inference of decision makers' judgments, and it is easy to obtain $c_{12} = 3$, $c_{13} = 9$, $c_{23} = 3$. Then, estimating $P_{4.1}$ by taking the restraint condition of Formulas (4.6)–(4.7) into consideration, one can obtain $w_1 = 0.632$, $w_2 = 0.211$, $w_3 = 0.158$, and the reverse inference result is $c_{12} = 3$, $c_{13} = 4$, $c_{23} = 1.4$. One can find that the latter c_{13} is too far away from the former c_{13} (the former value is 9 and the latter is 4). The result holding with Formula (4.6)–(4.7) is closer to the decision makers' original judgment.

After estimating $P_{4.1}$, one can obtain the optimal value c^* and a set of weight w_i. Then, rank w_i to obtain the ultimate order of the alternatives.

4.3 Aggregation of Two Kinds of Three-Point Interval Number Preferences

Referring to the definition of the three-point interval number reciprocal comparison matrix, the definition of the three-point interval number complementary comparison matrix will be put forward.

Definition 4.4

The matrix $\overline{B} = (\overline{b}_{ij})_{n \times n}$ is called the interval number complementary comparison matrix, where $\overline{b}_{ij} = [b_{ij}^L, b_{ij}^U]$, $b_{ij}^L \le b_{ij}^U$, $\overline{b}_{ji} = [1 - b_{ij}^U, 1 - b_{ij}^L]$, $\overline{b}_{ii} = [0.5, 0.5]$.

Definition 4.5

The matrix $\overline{D} = (\overline{d}_{ij})_{n \times n}$ is called the three-point interval number complementary comparison matrix, where $\overline{d}_{ij} = [d_{ij}^L, d_{ij}^M, d_{ij}^U]$, $d_{ij}^L \le d_{ij}^M \le d_{ij}^U$, $\overline{d}_{ji} = [1 - d_{ij}^U, 1 - d_{ij}^M, 1 - d_{ij}^L]$, $\overline{d}_{ji} = [0.5, 0.5, 0.5]$.

Suppose that decision makers adopt two kinds of uncertain comparison matrices to express their preference in group decision-making process. How to aggregate two kinds of uncertain preference effectively will be discussed in this part. The decision maker $i = 1, \dots, m$ provides the three-point interval number reciprocal comparison matrix. One can record these comparison matrices as the set

I and record the three-point interval number reciprocal comparison matrix as C_i, $i \in I$. The decision maker $i = m + 1, \ldots, n$ provides the three-point interval number complementary comparison matrix. One can record those as the set J and record the three-point interval number complementary comparison matrix as $D_j, j \in J$.

If some decision maker adopts the two-point interval number reciprocal comparison matrix or complementary comparison matrix, then one can transform it into three-point interval number reciprocal comparison matrix A' and three-point interval number complementary comparison matrix B', where $A' = ([a_{ij}^L, a_{ij}^M, a_{ij}^U])_{n \times n}$, $B' = ([b_{ij}^L, b_{ij}^M, b_{ij}^U])_{n \times n}$, $a_{ij}^M = (a_{ij}^L + a_{ij}^U)/2$, $b_{ij}^M = (b_{ij}^L + b_{ij}^U)/2$. According to C_i, $i \in I$, every comparison of its three-point interval number $[c_{ij}^L, c_{ij}^M, c_{ij}^U]$ c_{ij}^L (or c_{ij}^M, c_{ij}^U) can be aggregated by the ordered weighted averaging (OWA) method. Then, one can obtain the three-point interval number reciprocal comparison matrix C. Handle $D_j, j \in J$ using the same method and obtain the matrix D.

If decision makers adopt the three-point interval number reciprocal comparison matrix $C = [c_{ij}^L, c_{ij}^M, c_{ij}^U]_{n \times n}$ to express their preference, c_{ij}^M indicates the most possible judgment, and c_{ij}^L, c_{ij}^U are the lower and upper value. One can record the set of wc_i as the weight of the three-point interval number reciprocal comparison matrix. If it is completely consistent, Formulas (4.1)–(4.3) are satisfied.

If decision makers adopt the three-point interval number complementary comparison matrix $D = [d_{ij}^L, d_{ij}^M, d_{ij}^U]_{n \times n}$ to express their preference, one can introduce the error value dp_{ij}, dd_{ij} dpo_{ij}, ddo_{ij}. wd_i can be recorded as the weight. If it is completely consistent, Formula (4.10) is satisfied.

$$d_{ij}^L \leq d_{ij}^M = \frac{wd_i}{wd_i + wd_j} \leq d_{ij}^U \tag{4.10}$$

If D is completely consistent, to the error of

$$d_{ij}^M = \frac{wd_i}{wd_i + wd_j},$$

one can introduce the error variables dpo_{ij}, ddo_{ij}, and the relationship is $dpo_{ij} \geq 0$, $ddo_{ij} \geq 0$ and $dpo_{ij} \times ddo_{ij} = 0$. Formula (4.11) will be satisfied.

$$d_{ij}^M = (wd_i + wd_j) - wd_i + dpo_{ij} - ddo_{ij} = 0 \tag{4.11}$$

If Formula (4.10) does not satisfy

$$d_{ij}^L \leq \frac{wd_i}{wd_i + wd_j} \leq d_{ij}^U,$$

one can introduce the error variable dp_{ij}, dd_{ij}, and then Formula (4.12) satisfies.

$$\begin{cases} d_{ij}^L (wd_i + wd_j) \leq wd_i + dp_{ij} \\ wd_i \leq d_{ij}^U (wd_i + wd_j) + dd_{ij} \end{cases} \tag{4.12}$$

The lower the value of $\sum_{i,j} dpo_{ij} + ddo_{ij}$ is, the shorter the distance of deviation from the possible value d_{ij}^M is. The lower the value of $\sum_{i,j} dp_{ij} + dd_{ij}$ is, the shorter the distance of deviation from the upper–lower is. One can record the deviation distance value as $\sum_{i,j} dpo_{ij} + ddo_{ij} + dp_{ij} + dd_{ij} \to \min$. If $\sum_{i,j} dpo_{ij} + ddo_{ij} + dp_{ij} + dd_{ij} \to \min$, and it shows the better consistency of the three-point interval number complementary comparison matrix, model $P_{4.2}$ can be suggested to estimate the weight.

$$\min d = \sum_{i,j} dp_{ij} + dd_{ij} + dpo_{ij} + ddo_{ij}$$

$$\text{s.t.} \begin{cases} d_{ij}^M (wd_i + wd_j) - wd_i + dpo_{ij} - ddo_{ij} = 0 \\ d_{ij}^L (wd_i + wd_j) \leq wd_i + dp_{ij}, \ i, j = 1, \dots, n, \ i \neq j \\ wd_i \leq d_{ij}^U (wd_i + wd_j) + dd_{ij}, \ i, j = 1, \dots, n, \ i \neq j \\ \sum_{i=1}^n wd_i = 1 \\ wd_i \geq 0, \ dpo_{ij} ddo_{ij} = 0, \ dp_{ij}, \ dd_{ij}, \ dpo_{ij}, \ ddo_{ij} \geq 0, \ i, j = 1, \dots, n \end{cases} \quad (\text{P}_{4.2})$$

One can record the optimal value of model $P_{4.2}$ as d^* and obtain the following conclusions.

Theorem 4.5
If $d^* = 0$, the three-point interval number comparison matrix has complete consistency. Otherwise, it is also satisfied.

Proof
One can refer to the proof of Theorem 4.1.

Theorem 4.6

If $d^* > 0$, the three-point interval number comparison matrix $\overline{D}$ does not have complete consistency. The greater the value of d^* is, the worse the consistency of three-point interval number comparison matrix $\overline{D}$ is.

Proof

One can refer to the proof of Theorem 4.2.

Theorem 4.7

There must be an optimal value for model $P_{4.2}$.

Proof

One can refer to the proof of Theorem 4.3.

In the process of group decision making, decision makers should try to reach consistency, and then Formula (4.13) is satisfied.

$$wc_i = wd_i, \ i = 1, \ldots, n \tag{4.13}$$

Based on Formula (4.13), one can record $w_i = wc_i = wd_i, \ i = 1, \ldots, n$ and can join models $P_{4.1}$ and $P_{4.2}$ into $P_{4.3}$.

$$\min f = \sum_{i,j} (cp_{ij} + cd_{ij} + cpo_{ij} + cdo_{ij}) + \sum_{i,j} (dp_{ij} + dd_{ij} + dpo_{ij} + ddo_{ij})$$

$$\text{s.t.} \begin{cases} c_{ij}^M w_j - w_i + cpo_{ij} - cdo_{ij} = 0 \\ c_{ij}^L w_j \leq w_i + cp_{ij}, \ i, j = 1, \ldots, n, \ i \neq j \\ w_i \leq c_{ij}^U w_j + cd_{ij}, \ i, j = 1, \ldots, n, \ i \neq j \\ d_{ij}^M (w_i + w_j) - w_i + dpo_{ij} - ddo_{ij} = 0 \\ d_{ij}^L (w_i + w_j) \leq w_i + dp_{ij}, \ i, j = 1, \ldots, n, \ i \neq j \\ w_i \leq d_{ij}^U (w_i + w_j) + dd_{ij}, \ i, j = 1, \ldots, n, \ i \neq j \\ \sum_{i=1}^n w_i = 1, \ w_i \geq 0, \ i = 1, \ldots, n \\ cpo_{ij} cdo_{ij} = 0, \ dpo_{ij} ddo_{ij} = 0, \ cp_{ij}, \ cd_{ij}, \ cpo_{ij}, \ cdo_{ij}, \ dp_{ij}, \ dd_{ij}, \\ \qquad dpo_{ij}, \ ddo_{ij} \geq 0, \ i, j = 1, \ldots, n \end{cases} \tag{$P_{4.3}$}$$

The meaning of variables and formulas above model $P_{4.3}$ are the same for $P_{4.2}$ and $P_{4.1}$. One can record the optimal value of model $P_{4.3}$ as f^* and obtain the following conclusions.

Theorem 4.8

If $f^* = 0$, the group reaches complete consistency. Otherwise, it is also satisfied.

Proof

If $f^* = 0$, cp_{ij}, cd_{ij}, cpo_{ij}, cdo_{ij}, dp_{ij}, dd_{ij}, dpo_{ij}, $ddo_{ij} = 0$ is satisfied. There is a set of weights satisfying Formulas (4.2)–(4.3) and (4.11)–(4.12). This means that the group reaches complete consistency. If the group reaches consistency, it indicates there is a set of weights w_i, $i = 1, \ldots, n$ satisfying Formulas (4.2)–(4.3) and (4.11)–(4.12). As a result, the conclusion cp_{ij}, cd_{ij}, cpo_{ij}, cdo_{ij}, dp_{ij}, dd_{ij}, dpo_{ij}, $ddo_{ij} = 0$ can be obtained. That is, $f^* = 0$.

Theorem 4.9

If $f^* > 0$, the group does not reach consistency. The bigger f^* is, the worse the consistency is and the more dispersible the group opinion is.

Proof

If $f^* > 0$, one of the error values (or cp_{ij}, cd_{ij}, cpo_{ij}, cdo_{ij}, dp_{ij}, dd_{ij}, dpo_{ij}, $ddo_{ij} = 0$) must not be zero. The set of weights w_i, $i = 1, \ldots, n$ satisfying Formulas (4.2)–(4.3) and (4.11)–(4.12) does not exist. So the group does not reach complete consistency. The bigger f^* is, the bigger the deviation error is, and the more dispersible the group opinion is.

Theorem 4.10

There must be an optimal value of model $P_{4.3}$.

Proof

One can refer to the proof of Theorem 4.3.

One can estimate model $P_{4.3}$ to obtain f^* and w_i, $i = 1, \ldots, n$. One can check the consistency of group preference according to the value of f^* and analyze the alternative order via the w_i, $i = 1, \ldots, n$. In some situations, decision makers may want to know the uncertain characteristic concealed by the certain weight. In addition, w_i is a certain value while the preference of decision maker is uncertain. The certain weight cannot reflect the decision-making uncertainty preference. In this case, the weight distribution model with the error scope γ is suggested, where the f^* is obtained from $P_{4.3}$. The γ can be decided by decision makers. Generally, the lower γ is, the better the decision-making precision is.

$$\min/\max w_i$$

$$\text{s.t.} \begin{cases} \sum_{i,j} (cp_{ij} + cd_{ij} + cpo_{ij} + cdo_{ij}) \\ \quad + \sum_{i,j} (dp_{ij} + dd_{ij} + dpo_{ij} + ddo_{ij}) \le (1+\gamma) f^* \\ c_{ij}^M w_j - w_i + cpo_{ij} - cdo_{ij} = 0 \\ c_{ij}^L w_j \le w_i + cp_{ij}, \ i,j = 1, \dots, n, \ i \ne j \\ w_i \le c_{ij}^U w_j + cd_{ij}, \ i,j = 1, \dots, n, \ i \ne j \\ d_{ij}^M (w_i + w_j) - w_i + dpo_{ij} - ddo_{ij} = 0 \\ d_{ij}^L (w_i + w_j) \le w_i + dp_{ij}, \ i,j = 1, \dots, n, \ i \ne j \\ w_i \le d_{ij}^U (w_i + w_j) + dd_{ij}, \ i,j = 1, \dots, n, \ i \ne j \\ \sum_{i=1}^n w_i = 1, \ w_i \ge 0, \ i = 1, \dots, n \\ cpo_{ij} cdo_{ij} = 0, \ dpo_{ij} ddo_{ij} = 0, \ cp_{ij}, \ cd_{ij}, \ cpo_{ij}, \ cdo_{ij}, \ dp_{ij}, \ dd_{ij}, \\ \quad dpo_{ij}, \ ddo_{ij} \ge 0, \ i,j = 1, \dots, n \end{cases} \tag{$P_{4.4}$}$$

Theorem 4.11
There must be an optimal value of model $P_{4.4}$.

Proof
One can obtain f^* according to model $P_{4.3}$ and a set of w_i, $i = 1, \dots, n$. It must correspond with a group of estimation via model $P_{4.4}$. So there must be an optimal value of model $P_{4.4}$.

According to the above method, one can obtain the weight distribution via estimating the weight of the three-point interval number comparison matrix. If w_i, $i \in (1, \ldots, n)$ is unique, one can obtain the conclusion of min w_i = max w_i.

4.4 Base Paper Specification Reduction Based on Three-Point Numbers

4.4.1 Background and Problems

World corrugated base paper production increased from 92 million tons to 81 million tons from 1995 to 1999. In 2004, production was more than 110 million tons, especially in China where the production in 2004 reached 16.4 million tons, and the consumption of corrugated base paper in China is greater. There are many manufacturers and different quality levels of base paper in the paper market. In addition, each type of paper has different raw paper width and weight of the points, which results in a very large number of base paper specifications. For a certain number of carton manufacturing enterprises, in order to satisfy the needs of multistandard carton, they need to use 700 or 800 kinds of specifications of the base paper each month, and each type of average is close to 2000. If a company buys paper without reasonable plans based on clear specifications, the amount of unused stock stored in its warehouse will increase and ultimately the company will need to increase its storage capacity to solve the problem. With the rapid expansion of production capacity, the difficulty of paper warehouse management becomes more apparent. It not only increases the difficulty of management but also costs the enterprise a lot of money. Are there any scientific methods that can be used to reduce the amount of base paper without influencing the requirement of the company yet reduce the difficulty in the process of the management?

Actually, it is not a difficult task to reduce the base paper size appropriately; the key point is to assess the size of the base paper streamlined program so it can meet the procurement, finance, management, planning, marketing, and other requirements. According to the author's available information, there are no relevant research reports available. Due to the large number of paper sizes, the reasonable scheme selection requires a high workload; whether the schemes are reasonable or not can directly affect the business activities. How to screen the reasonable streamline programs of base paper? Using a computer simulation method is the direct way but the method should use system simulation tools to describe the operating procedure of the entire supply chain, and run many cycles according to parameters of the programs to comparatively analyze operational status and economic indicators based on the streamlined program, and then select the best one. Because it is difficult to set the process parameters of the enterprise operation, such a method is difficult to achieve. This section considers that the number of base paper specifications can rely on the experience of management staff; comprehensive materials; production, quality, and other management departments; or even appropriate advice from their suppliers and vendors; and the recommendations from their users to help the company obtain valuable experience and thus achieve a reasonably streamlined program.

As an example, consider a large enterprise producing corrugated cartons in the east area of China, which has all the plate-making, printing, and molding equipment. It experienced development from township enterprises, collective enterprise to private enterprise, so it had some problems such as base paper management efficiency being low, the division of rights and liabilities not being clear, and base paper storage and management level being the same as at the beginning. The warehouse crowding problem get worse with the market development having increased output. So the enterprise has to build more warehouses, but the problem is not solved completely. The main focus of this part is as follows:

1. Use AHP to streamline base paper size and design the evaluation index system for evaluation of streamlining programs.
2. Integrate judgments of the site management personnel by the three-point interval method, let site management experts use the multiple comparison method to determine the advantages and disadvantages of streamline programs, and rank the programs according to score to obtain the satisfactory program.

4.4.2 Criteria for Base Paper Specification Reduction

Through communication with technicians and managers, the impact of specifications of the base paper on business in all aspects is analyzed,

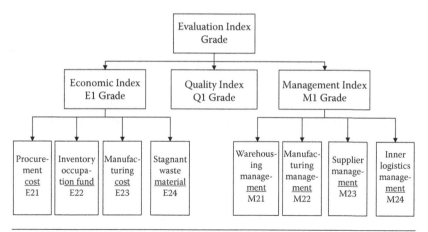

Figure 4.3 Base paper specification streamline criteria index.

which can be used to establish evaluation criteria in terms of economy, management, and quality (see Figure 4.3).

The economic index from the financial perspective is to analyze the paper specification streamlined program for the production, management costs, and other aspects, mainly on the procurement costs, inventory and occupation funds, manufacturing costs, and stagnant waste material. Procurement costs refer to activities of procurement bidding, negotiation, ordering, transportation, and other expenses incurred by related activities combined. In general, after simplification of the base paper* streamlined specifications, the same amount of base paper is purchased, but use of a single type saves procurement costs. The inventory occupation fund is the total amount of corporate liquidity occupied by base paper.

Streamlined base paper specifications and reasonable base paper specifications will be more effective, and operate a reasonable production plan, greatly reducing the inventory. The manufacturing cost is the sum of the corrugated plates, molding, printing, and packaging consumed in the process of labor; fuel and other materials; and equipment wear and tear. The appropriate base paper specifications can be adjusted to reduce the production time and preparation and equipment costs, simplify production processes, and thus reduce manufacturing costs appropriately. Stagnant

---

* Base paper can be processed into various kinds of paper products.

waste material is unused base paper; waste material refers to excess materials and waste. Streamlining the process must meet specifications of the substitute poor quality for good quality and substitute good quality for poor quality issue. There may also be a need to communicate with the user, who permits appropriate adjustments to the contract, the contract compensation for losses, and waste reduction costs.

The management index analyzes the influence of base paper size specification for management workload, labor complexity, and management efficiency. The management index is composed of four secondary indexes such as warehouse management, production management, supplier management, and internal logistics management. Warehouse management includes in–out and inventory warehouse for base paper; production management refers to optimization, plan, organize implementation, and control management work of the production process; supplier management mainly involves supplier evaluation and establishing partnerships; internal logistics mainly involves the base paper feeding, retreating material, and waste disposal work (only inspection of the internal logistics activities that directly relate to the storage management).

Quality index is the analysis the contribution of streamlined base paper for the improvement of product quality. Proper base paper size number would reduce production error probability. Production preparation and adjusting work would benefit the quality of the product to some extent.

Base paper's basic characteristics include brand name (e.g., JiuLong, XingBao, etc.), quality grades (such as class A, class B, and class C), purpose (e.g., flour paper, glossy paper, watts paper), and paper gram weight and width (paper roll height). Quality grades and purpose are few, so no replacement is possible among the different quality grades and purposes. But brand name, gram weight, and width grades have hundreds of variations. Therefore, the base paper specification mainly referred to brand name, gram weight, and width in this section.

Streamline by changing the level of the minimum width. Paper width of the current enterprise of the minimum width of the minimum level (paper produced by base paper suppliers) grade is 5 cm. Production process needs to cut the edge of base paper. For instance, to produce paper 200 cm wide, 205 cm width of the base paper must be used. But there is a relevant provision of the maximum cutting edge, approved under special circumstances to allow expansion of the scope of the conference side. A larger than necessary proportion of the paper is waste. There are three options—program A1: kept to a minimum width of grade to 5 cm; program A2: change the level of the minimum width to 10 cm; program A3: change rating minimum width to 15 cm.

Streamline by changing the level of the minimum weight. The paper weight of the current enterprise of the minimum weight of the minimum level grade is 5 g/m^2; there is also special paper weight grade, such as the weight grade of 112 g/m^2 of Guiheiwa. Special paper can be used to downgrade or upgrade in accordance with the relevant provisions. The paper production plan must be in accordance with the weight of paper used under the normal production conditions. Paper specifications may be downgraded or upgraded only under special situations (usually only downgraded), but it must ensure that the quality is not lower than the specified product quality. There are three options— program B1: to maintain the original weight class; program B2: to maintain the original weight class but remove the special gram specifications; program B3: change the minimum weight class to 10 g/m^2.

Streamline the paper size by reducing the number of suppliers. From the current average, enterprise purchases the base paper from about 80 vendors of base paper suppliers and changes every year. Twenty of them were designated as important suppliers. There are three options—program C1: 20 key suppliers, reduce the number of suppliers; program C2: 15 key suppliers, reduce the number of suppliers; program C3: 10 key suppliers, reduce the number of suppliers.

4.4.3 Preference and Result Analysis

The evaluation index system of base paper specifications includes first grade indexes E1, M1, and Q1, as well as secondary indicators, needed to determine the priority weight of each index factor. This section scale is based on 0.1–0.9 [57] on-site experts to judge between two methods to determine the priority weight.

Table 4.1   Results of the Relative Importance of Indicators after Finishing the Questionnaire

E1:M1	0.6	E21:E22	0.4	E22:E23	0.7	M21:M22	0.4	M22:M23	0.7
E1:Q1	0.6	E21:E23	0.3	E22:E24	0.7	M21:M23	0.7	M22:M24	0.6
M1:Q1	0.7	E21:E24	0.6	E23:E24	0.5	M21:M24	0.3	M23:M24	0.4

The questionnaires about the problem are distributed to the purchasing department, production department, quality assurance, and other related departments to obtain data about comparisons and alternatives. Department managers, business managers, technical directors, general workers, and industrial engineers participate in the survey. The data processing results of the questionnaires are shown in Table 4.1.

To integrate multiple suggestions and accurately reflect the consolidated results of questionnaire data, many methods can be used, such as the average method, the weighted average method, and interval methods. If using the average method,

$$\alpha_j = \sum_{i=1}^{m} a_{ij}/m,$$

m is the number of the site management participating in the survey. The disadvantage is the same view of data from different investigators; it does not reflect the difference of knowledge or individual ability. If using the weighted average method,

$$\alpha_j = \sum_{i=1}^{m} a_{ij}k_i,$$

k_i is the priorities of the decision makers. The disadvantage is that consolidated results do not reflect the different views between decision makers or the possibility of missing values within the range of different information. If using the method of interval number method to accommodate multiple views, the range must get too large in order to cover the whole, resulting in an increase of the degree of uncertainty in decision making, and decision makers cannot get the range to determine the distribution. Based on existing research in this section, a three-point interval number method is introduced, which not only maintains the range of the interval but also highlights the most likely value of the focus point. It can be a more extensive and accurate expression of the preference of decision makers, to a certain extent making up for the shortage of interval end points. Twenty survey questionnaires were distributed (specific data omitted), with three-point interval number methods that decision makers prefer to get the following results:

1. **Streamline evaluation index weight of base paper specifications:** managers to be judged on three aspects: first degree indexes of economic, management, and quality. Change complementary judgment. The following three-point interval number comparison matrix is obtained:

$$
\begin{bmatrix}
[1, 1, 1] & [0.5, 1, 2] & [0.5, 1.3, 3] \\
[0.5, 1, 2] & [1, 1, 1] & [0.333, 1.3, 2] \\
[0.333, 0.8, 2] & [0.5, 0.8, 3] & [1, 1, 1]
\end{bmatrix}.
$$

Based on model $P_{4.1}$, the weight result is [E1, M1, Q1] = [0.367, 0.359, 0.274], objective function value $c^* = 0.034$. The general comments of site managers are as follows: The primary consideration of the base paper specifications to be streamlined is economic index, which also reflects the effectiveness of the supremacy in business, then consider the feasibility and convenience by streamlining the management and operation. Paper size cannot work to streamline the expense of efficiency.

Four secondary indexes are carried out on the priority of importance under the economic index:

$$
\begin{bmatrix}
[1, 1, 1] & [0.5, 0.8, 1] & [0.5, 1.1, 2] & [0.5, 1.2, 3] \\
[1, 1.25, 2] & [1, 1, 1] & [1, 1.3, 2] & [1, 1.5, 3] \\
[0.5, 1, 2] & [0.5, 0.8, 1] & [1, 1, 1] & [0.5, 1.2, 2] \\
[0.333, 0.83, 2] & [0.333, 0.667, 1] & [0.5, 0.83, 2] & [1, 1, 1]
\end{bmatrix},
$$

weight result is [E21, E22, E23, E24] = [0.247, 0.309, 0.237, 0.206], $c^* = 0.0238$. This shows that management gives priority to inventory costs and lag to material cost factors. The matrix is based on the priority of four secondary indicators of management indexes:

$$
\begin{bmatrix}
[1, 1, 1] & [0.5, 0.8, 2] & [1, 1.2, 3] & [0.5, 0.9, 1] \\
[0.5, 1.25, 2] & [1, 1, 1] & [1, 1.5, 2] & [0.8, 1.1, 2] \\
[0.333, 0.83, 1] & [0.5, 0.667, 1] & [1, 1, 1] & [0.5, 0.8, 1] \\
[1, 1.11, 2] & [0.5, 0.9, 1.25] & [1, 1.25, 2] & [1, 1, 1]
\end{bmatrix},
$$

and the result is [M21, M22, M23, M24] =[0.231, 0.298, 0.199, 0.272], $c^* = 0.0464$. There is more emphasis on convenience for management at site manufacturing after the specifications of the base paper are reduced, while the impact on the importance of management of supplier is the least important. Using the combination weight method to get the value of combination weight [E21, E22, E23, E24, M21, M22, M23, M24, Q1] = [0.090, 0.112, 0.088, 0.077, 0.083, 0.107, 0.071, 0.098, 0.274]. In all evaluations, the importance of quality index shares the largest weight value of 0.274, followed by the inventory occupation fund index, the weight value reached to 0.112.

2. **Survey and score of base paper specification streamlined program:** Code names A1–3, B1–3, C1–3 correspond to the streamlined program portfolio of base paper specification A, B, C, corresponding to E21–24, M21–24, Q1 in order to streamline the evaluation of specifications for the base paper index system. Each streamlined program's contribution to the grade project is considered, scoring points from 0 to 100 points. The survey object is expanded based on an evaluation questionnaire, including materials, production, quality control department, finance, customer management, sales and marketing ministry of personnel, and other related departments. Fifty questionnaires were issued; the score is obtained by the following formula:

$$
w_i = \sum_{j=1}^{9} \alpha_j \times p_{ij}
$$

In the formula, p_{ij} is the score about the jth index of the ith program; α_j is the weight coefficient of the jth program; w_i is the final score of the program. The final score of all

Table 4.2   Priority Score of Base Paper Specification Streamlined Program

	PROGRAM	NAME OF PROGRAM	SCORE
A	A1	Kept to a minimum width of grade to 5 cm	73.8
	A2	Change the level of the minimum width to 10 cm	85.1
	A3	Change the level of the minimum width to 15 cm	73.3
B	B1	Maintain the original weight class	81.4
	B2	Maintain the original weight class but remove the special gram specifications	82.2
	B3	Change the minimum weight class to 10 g/m²	69.7
C	C1	20 Key suppliers, reduce the number of suppliers	80.5
	C2	15 Key suppliers, reduce the number of suppliers	77.2
	C3	10 Key suppliers, reduce the number of suppliers	72.2

surveys for each program is w; the average $\overline{w}_i$ is the score of each program as shown in Table 4.2.

Program A2 is chosen from Group A, namely, changing the level of the minimum width to 10 cm; program B2 is chosen from Group B, namely, maintaining the original weight class of 5 g/m², not using nonstandard base paper weight; program C2 is chosen from Group C, namely, maintaining the original weight class but removing the special gram specifications.

3. **Review results of base paper specification streamlining:** Base paper specification streamlining is evaluated in three aspects: width grade range, weight class range, and the number of suppliers. A streamlined portfolio in paper specification is obtained: A2 + B2 + C1. Program A2 will reduce the demand for base paper by half and the effect is obvious. Program B2 has little effect but is also the easiest to execute. Program C1 can only be a qualitative improvement. Minimizing the number of suppliers is the key to improving the efficiency of inventory management. C3 is more attractive compared to C2, but the scores of C2 and C3 were lower. The number of suppliers at this stage is generally considered to not be reduced substantially. If it is forced to reduce, it will result in an increase of waste. Even if the inventory management and the cost of purchasing raw paper declines, in the end it may also increase the total cost, so the score was consciously controlled in the process.

Taking into account the effect of program and effectiveness, A2 and B2 were proposed to be implemented first. Take December 2005 as an example. A weight that is not 0 or 5 should be adjusted to a higher grade; 5 cm width at the end of the specification is uniform up to 10 cm into the end of the specification; the base paper is only slightly larger than the half of the total amount of the original specifications, so economic and management efficiency will be improved significantly.

Paper management is critical; it is necessary to strengthen the management of base paper, quality, storage, use, and other aspects of paper management. The differences of base paper use and place of origin lead to a huge number of paper specifications, and lack of reasonable base paper specifications, so the amount of warehouse safety stock and unused products rises. Then the difficulty of management increases and more costs are incurred. How to effectively control the paper size and paper specifications to streamline the screening program is to reasonably manage difficult issues. This section uses an analytic hierarchy to streamline paper specification, assembling the preference of experts based on the three-point interval number. The application has a strong operational simplicity. The focus of future research should be to further verify the paper size to streamline the program, strengthen the three-point interval number AHP theory and applied research, and theoretically improve the three-point interval number AHP consistency test and the weight estimation methods.

4.5 Summary and Future Research

As decision-making problems are increasingly complex, more and more people join the decision-making process, and group decision making is widely used. Decision makers may adopt different kinds of preferences, and it is usually unrealistic to depict complex problems via the definite preference in some situations. Therefore, the three-point interval number comparison matrix is suggested. The consistency and weight estimation based on the three-point interval number comparison matrix are studied. Some theory questions should be studied further, such as other effective approaches to weight estimation. The questions to be studied further can be summarized as follows:

1. This chapter defines the most likely value. In fact, in some complex decision scenarios, the views are not very concentrated in decision-making situations. The most likely value of the definition is complex; it is possible that the most likely value does not have single value and requires further study for reasonable application.

2. Compared with the two end points of interval number, the number of three-point intervals increases the amount of information. However, how to effectively build the two end points and three-point interval number preferences is still worth considering. On the one hand, if the two end points are changed into the three-point interval number, additional information is uncertain. On the other hand, if the three end points are changed into the second end point, the lost information is the most important decision information. This information may result in loss of decision making of the series of problems.

3. There are several ways to transform the information, such as interval numbers, fuzzy numbers, and linguistic variables. Ways to transform the three end point preferences and other preferences need to be studied. The transformation of three end point references may be more convenient for this conversion. Further research is needed.

4. In addition, similar to the two end points interval number, the problem is the internal differences and suitability of reciprocal judgment preference and complementary preference information. Despite some methods for the conversion about two kinds of information, certain information will be lost in the process of conversion on different preference. How to measure information is a difficult problem. In a sense, no simple mathematical method can solve it. It may need application group psychology and behavior decision theory.

5

DECISION-MAKING METHOD OF THE LINGUISTIC COMPARISON MATRIX

5.1 Introduction

Due to the complexity of the alternatives and human thinking process, the linguistic label is usually adopted, which concerns the consistency of linguistic judgment, the transformation of different linguistic scales, and the decision-making method in group and in the incomplete information environment [108–111]. In recent years, many scholars have focused on the research of the linguistic comparison matrix, especially the integration method, the consistency measure, and the consistency improvement of linguistic comparison matrix. The internal consistency characteristics of the linguistic comparison matrix reflect the decision maker's rationality in logical judgment, which plays a vital role in scientific decision making. Two aspects are studied in this chapter the satisfaction consistency of the linguistic comparison matrix and the ordinal consistency of the linguistic comparison matrix. With respect to satisfaction consistency in the linguistic comparison matrix, the definition of strength in complete consistency of the linguistic comparison matrix is given. But it is difficult to find the completely consistent comparison matrix in the process of decision making. So what are the conditions that the matrix should satisfy to meet the consistent linguistic preference relation? The matrix educed from the linguistic comparison matrix is used to research the consistency. But the scaling of the educed matrix is different from the traditional reciprocal comparison matrix, so it cannot determine whether the linguistic comparison matrix meets the consistent linguistic preference relation according to the consistency ratio of the educed matrix. The conclusions of the existing literature [100, 109] have the reference values to measure the satisfaction consistency of the linguistic comparison matrix. With respect to transitive linguistic preference relation, the practicable method to check the ordinal consistency of linguistic comparison matrix is proposed in the literature. However, there is still no literature that provides a method to improve the linguistic comparison matrix, which does not meet ordinal consistency. The consistent problem of fuzzy comparison matrix is studied in the literature (the scale of elements in the comparison matrix is 0.1 to 0.9) [91, 100, 101, 108, 109]. There are some conclusions of comparison matrix in the literature [108–111] which give much enlightenment for the research in this chapter. Based on the existing research, the tasks of this chapter are to: (1) propose a new measure index for the satisfaction consistency of the linguistic comparison matrix and propose a method to improve matrix that does not meet the consistent linguistic preference relation; (2) propose a method to determine the ordinal consistency and propose heuristic modification rules for the matrix that does not meet the transitive linguistic preference relation.

5.2 Consistency Measurement and Modification Method of Linguistic Comparison Matrix

5.2.1 Basis Definition of the Linguistic Preference

There are two expressions of the linguistic scale of linguistic comparison matrix.

1. Define the language phrase set. $S = \{s_a | a = 0, \ldots, 2g\}$ a is integer, and s_a expresses that these alternatives are good or bad in multiple comparison. S is described as the language phrase set. If $g = 6$, s includes 13 elements. $S = \{s_0$ = absolutely poor, s_1 = extremely poor, s_2 = very poor, s_3 = poor, s_4 = poorer, s_5 = slightly poor, s_6 = fair, s_7 = slightly good, s_8 = better, s_9 = good, s_{10} = very good, s_{11} = extremely good, s_{12} = absolutely good$\}$.

2. Define the language phrase set. $S = \{s_{a'} | a = -t, \ldots, 0, \ldots, a$ is an integer. S includes 13 elements: $S = \{s_{-6}$ = absolutely poor, s_{-5} = extremely poor, s_{-4} = very poor, s_{-3} = poor, s_{-2} = poorer, s_{-1} = slightly poor, s_0 = fair, s_1 = slightly good, s_2 = better, s_3 = good, s_4 = very good, s_5 = extremely good, s_6 = absolutely good$\}$.

There are no essential distinctions in consistency measurement of these two language phrase sets. The two linguistic models can be transformed according to $s_\alpha = s_{\alpha'+g} = s_{\alpha'+6}$, and this chapter uses the second one.

Suppose that a decision maker gives multiple comparisons according to the definition of S. Construct the linguistic comparison matrix $A = (a_{ij})_{n \times n}$. There are $a_{ii} = s_0; a_{ij} = s_\alpha \in S, a_{ji} = s_{-\alpha}, i, j = 1, \ldots, n$.

5.2.2 Definition of the Satisfaction Consistency of the Linguistic Comparison Matrix

Definition 5.1

Suppose $S = \{s_i \mid i \in [-t, t]\}$, $s_i \in S$, i according to I

$$I : S \to N$$
$$I(s_i) = i, s_i \in S$$

(5.1)

For $i, j \in [-t, t]$, if s_i is better than s_j, $i > j$ can be obtained and vice versa.

Definition 5.2

if $I(a_{ij}) = I(a_{ik}) - I(a_{jk})$ is met, the linguistic comparison matrix is completely consistent.

Because of the complexity during the decision-making process, it is too difficult to obtain the judgment of complete consistency. When can the linguistic judgment matrix of the decision maker yield a consistent judgment? The method for measuring linguistic preference is studied and the critical standard is proposed, which can be seen in Dong et al. [100, 101]. It can be summarized as follows: $\varepsilon_{ij} = I(a_{ij}) - \left[\sum_{k=1}^{n} I(a_{ik}) + I(a_{kj})\right] / n$ is regarded as the independent variable obeying the normal distribution. The average value equals 0 and the variance is δ. Actually, there is a strong relationship among a_{ij}, a_{ik}, a_{kj} in the completely consistent situation of $I(a_{ij}) = I(a_{ik}) - I(a_{jk})$. Does ε_{ij} obey independent normal distribution? Supposing the linguistic scale set is $[s_{-4}, \ldots, s_4]$, the procedure presented shows in Figure 5.1 whether ε_{ij} obeys the normal distribution.

In Figure 5.1, matrix A is a random matrix whose element is between −4 and 4. Based on this random matrix, matrix B can be

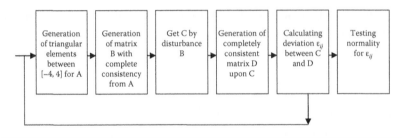

Figure 5.1   Idea of testing of random deviation normality.

obtained [100], where $b_{ij} = \left(\Sigma_{k=1}^{n}a_{ik} + a_{kj}\right)/n$ is completely consistent. Because matrix A is random, its consistency may be poor. Matrix C can be obtained by disturbing matrix B; $c_{ij} = b_{ij}(1+\Delta), \Delta$ is in $[0, 1]$. Then matrix D can be obtained based on the complete consistency, $d_{ij} = \left(\Sigma_{k=1}^{n}c_{ik} + c_{kj}\right)/n$. The deviation between matrices C and D can be calculated as ε_{ij}. The normality examination can be carried on after generating many ε_{ij}.

The Jarque-Bera method is used to test whether the sequence obeys the normal distribution. If the output result shows $h = 0$, it will not reject the hypothesis of normal distribution at a significance level 0.05. The hypothesis of normal distribution assumption will be rejected if $h = 1$. The value of p is the probability of statistics that are over the original observation. The distribution hypothesis assumption will be rejected when p is small. Many experiments based on the Jarque-Bera method are covered in this chapter, and one of the experimental results is shown in Figures 5.2-1 and 5.2-2.

The experimental results show that ε_{ij} does not meet the normal distribution. Therefore, the method proposed by Dong et al. [100] is worth deliberating. To solve the problems, a new measuring method is proposed in this chapter. If matrix A is completely consistent, according to Definition 5.2, $\forall i, j, k, I(a_{ij}) = I(a_{ik}) - I(a_{jk})$, namely, $I(a_{ij}) - I(a_{ik}) + I(a_{jk}) = 0$, thus $\Sigma_{i,j,k}\left|I(a_{ij}) - I(a_{ik}) + I(a_{jk})\right| = 0$. If it is not completely consistent, the relationship of $\Sigma_{i,j,k}\left|I(a_{ij}) - I(a_{ik}) + I(a_{jk})\right| \neq 0$ is obtained. The poorer the consistency of the linguistic comparison matrix is, the larger the differences between $\Sigma_{i,j,k}\left|I(a_{ij}) - I(a_{ik}) + I(a_{jk})\right|$ and 0 are. Formula (5.2) can be obtained to express the consistency of the linguistic comparison matrix.

$$\rho = \frac{\displaystyle\sum_{1 \leq i < j < k \leq n} \left|I(a_{ij}) - I(a_{ik}) + I(a_{jk})\right|}{C_n^3} \tag{5.2}$$

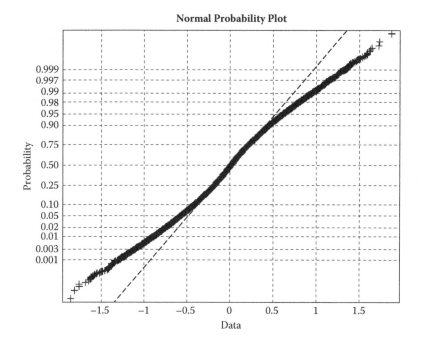

Figure 5.2-1 Normality text when $n = 4$ ($h = 1$, $p = 0$).

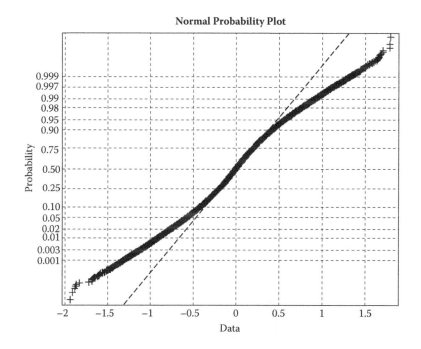

Figure 5.2-2 Normality text when $n = 5$ ($h = 1$, $p = 0$).

C_n^3 means that if there are N alternatives in multiple comparisons, there will be C_n^3 circuits taking every scheme as a node. For example, if there are four alternatives, four circuits can be obtained after comparison, 1-2-3, 1-2-4, 1-3-4, 2-3-4. If there are five alternatives, there will be 10 circuits, 1-2-3, 1-2-4, 1-2-5, 1-3-4, 1-3-5, 1-4-5, 2-3-4, 2-3-5, 2-4-5, 3-4-5. By the equation of ρ, its essence is the deviational meaning of every circuit's judgment.

Theorem 5.1

$\rho = 0$ is the necessary and sufficient condition of a completely consistent linguistic comparison matrix.

Proof

If $\rho = 0$, each part of absolute expression must be 0 due to $\rho = |\ | + \cdots + |\ |$ ("$|\ |$"means the operation rule of absolute value). It can be concluded that the matrix is completely consistent. Conversely, if the matrix is completely consistent, each part of the absolute expression must be 0, and the sum of the absolute expression also equals 0.

For the comparison matrix that is completely inconsistent, $\rho > 0$ can be obtained. The greater the value of ρ, the poorer the consistency is. When the value of ρ is more than a certain value, it can be considered that the judgment consistency of decision makers is very poor. In order to measure the consistency of the linguistic comparison matrix exactly, the linear membership function is established.

Definition 5.3

Suppose μ is the satisfaction consistency index of the linguistic comparison matrix,

$$\mu(\rho) = \begin{cases} 1 & \rho < x_1 \\ \rho \geq x_2, \mu(\rho) = \phi, x_1 \leq \rho < x_2 \\ \rho \geq x_2, \mu(\rho) = \phi, \rho \geq x_2 \end{cases} \tag{5.3}$$

In Formula (5.3), $x_2 > x_1 \geq 0$ are the allowable deviations of decisions that are set by the decision makers in advance. If the limitation of allowable deviation is seriously violated, $\rho \geq x_2$, $\mu \leq 0$ can be obtained. If the limitation of allowable deviation is completely met, the relationship of $\mu = 1$ can be obtained. If the limitation of allowable deviation is changed from a serious violation to completely met, the value of μ also increases monotonously from 0 to 1. Generally, when $\mu \geq 60\%$, it can be considered that the linguistic comparison matrix is consistent.

The method of setting x_1, x_2 is explained as follows. Suppose all elements are known except a_{ij} to estimate the value of a_{ij}. It is easy to obtain some circuits including a_{ij} $\left| I(a_{ij}) - (I(a_{ik}) - I(a_{jk})) \right| = \rho' \Rightarrow I(a_{ij}) = I(a_{ik}) - I(a_{jk}) \pm \rho'$.

If the decision makers' true preferences are s_x, they should answer the following two questions:

1. How large a ρ' can be considered the judgment value to reflect the decision makers' true approximate preferences? This is called the *true approximate preference region*.
2. How large a ρ' can the decision maker consider a_{ij} has not already been approximate preferences? This is called the *nonapproximate preference region*. The approximate preference region is between the true approximate preference region and the nonapproximate preference region (Figure 5.3).

Suppose that the decision maker's approximate preference region is bilateral symmetry. For example, if the true preference that results from the comparison of two alternatives is s_2 (Figure 5.4), the decision maker could consider that $a_{ij} \in [s_{1.5}, s_{2.5}]$ is also the true

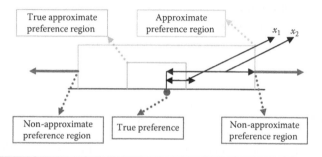

Figure 5.3 Schematic diagram of allowable deviation set of the decision.

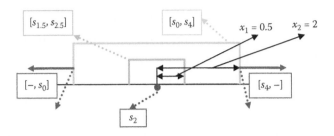

Figure 5.4 Schematic diagram of allowable deviation set of the decision.

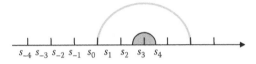

Figure 5.5 Allowable deviation set of decision based on scale.

preference. If $\rho' = 0.5$, $x_1 = 0.5$; if $a_{ij} \leq s_0$ or $a_{ij} \geq s_4$, it cannot be seen as the approximate preference if $\rho' = 0.2$, $x_2 = 2$.

Actually, according to the linguistic scale, which includes nine elements (Figure 5.5), if the true preference is s_1, $x_2 = 2$, then $[s_{-1}, s_3]$ can reflect the preference of decision maker. When $x_1 = 0.5$, the preference $[s_{0.5}, s_{1.5}]$ is not different from the true preference of s_1. It can be inferred that the linguistic set includes nine elements and $x_2 = 1.2$ can be obtained as in Xu [129]. It can be inferred that $x_2 = 2$ as in Song and Yang [130]. Generally, $x_2 \leq 2$ can be obtained based on intuitive understanding of the judgment.

If $x_1 = 0.5$, $x_2 = 2$, $\mu = 0.6$, $\rho = \mu(x_1 - x_2) + x_2 = 1.1$ can be obtained. This means that every circuit's average deviation value of acceptable judgment is 1.1. Is there any relationship between x_1, x_2's value setting and n? Actually, the accuracy is different in various decision problems; the number n is different in various compared alternatives. The decision makers could make suitable modification according to their preferences, so the relation between between x_1, x_2 and n is not discussed in this chapter.

5.2.3 Modification Method of Linguistic Comparison Matrix with Satisfaction Consistency

When the linguistic comparison matrix does not meet satisfaction consistency, how do we determine inconsistent elements in the linguistic comparison matrix? Two ways are proposed in Dong et al. [100]. The method proposed in this chapter is easier to improve consistency than the measurement scaling of satisfaction consistency. Suppose $A = (a_{ij})_{n \times n}$ cannot meet the condition satisfaction consistency. Then the decision maker adjusts only one element every time and keeps the others invariant.

Suppose $I(a_{st}) = x$, the problem is to solve

$$\max \mu(x)$$
$$s.t. x > 0$$

The equivalent model is

$$\min \rho(x) \tag{5.4}$$
$$s.t. x > 0$$

The expressions of $\mu(x)$ and $\rho(x)$ can be found in Formulas (5.3) and (5.2).

Theorem 5.2

$\min \rho(x)$'s expression can be simplified as follows, namely, $\min \rho \Rightarrow \min |x - x_1| + |x - x_2| + \cdots + |x - x_m|$, where, $x_1, \ldots, x_m$ are constants, x is an unknown variable, and $m = n - 2$.

Proof

By the expression of $\rho(x)$, Formula (5.5) can be obtained.

$$\min \rho(x) = \min \sum_{\substack{1 \le i < j < k \le n \\ i=s, j=t \text{ or } i=s, k=t \text{ or } j=s, k=t}} \left| I(a_{ij}) - I(a_{ik}) + I(a_{jk}) \right|$$

$$+ \sum_{\substack{1 \le i < j < k \le n \\ i \ne s, j \ne t \text{ or } i \ne s, k \ne t \text{ or } j \ne s, k \ne t}} \left| I(a_{ij}) - I(a_{ik}) + I(a_{jk}) \right| \tag{5.5}$$

The second term in the formula is unrelated to x, so it is a constant. Formula (5.5) can be simplified in the following form, namely,

$$\min \rho(x) = \min \sum_{\substack{1 \le i < j < k \le n \\ i=s, j=t \text{ or } i=s, k=t \text{ or } j=s, k=t}} \left| I(a_{ij}) - I(a_{ik}) + I(a_{jk}) \right| \tag{5.6}$$

It is easy to obtain

$$\min \rho(x) = \min \sum_{\substack{k=1 \\ k \neq s, k \neq t}}^{n} |x - x_k|.$$

Namely, $\min |x - x_1| + \cdots + |x - x_m|$, m is $n - 2$ nodes except s, t. $m = n - 2$. $x_f = I(a_{sk}) - I(a_{tk})$, f, $k = 1, \ldots, n$, $k \neq s, k \neq t$.

Theorem 5.3

Let $x_1, \ldots, x_m$ be a sequence from low value to high; if m is an odd number, the $\min \rho(x)$'s optimal solution is $x^* = x_{(m+1)/2}$. If m is an even number, the $\min \rho(x)$'s optimal solution is $x^* \in (x_{m/2}, x_{m/2+1})$. Moreover, to the left of x^*, $\rho(x)$ is a monotone decreasing function; to the right of x^*, $\rho(x)$ is a monotone increasing function.

Proof

Suppose $f(x) = |x - x_1| + |x - x_2| + \ldots + |x - x_l| + |x - x_{l+1}| + \ldots + |x - x_m|$, $x_l \leq x \leq x_{l+1}$ (l is indeterminate), the following formula can be obtained.

$$f(x) = (x - x_1) + \ldots + (x - x_l) + [-(x - x_{l+1}) - \ldots - (x - x_m)]$$

$$= lx - (x_1 + \ldots + x_l) + (x_{l+1} + \ldots + x_m) - (m - l)x$$

$$= (2l - m)x - (x_1 + \ldots + x_l) + (x_{l+1} + \ldots + x_m)$$

Suppose

$$\frac{df(x)}{dx} = 0 \Rightarrow 2l - m = 0,$$

l and m are integers, so

$$l = \left[\frac{m}{2}\right].$$

When m is an odd number, $l = \dfrac{m+1}{2}$, and

$$x^* = x_{\frac{m+1}{2}},$$

when m is an even number,

$$l = \frac{m}{2},$$

and $x^* \in (x_{m/2}, x_{m/2+1})$.

If

$$l < \frac{m+1}{2}, \quad \frac{df(x)}{dx} < 0$$

can be obtained when m is an odd number. If

$$l > \frac{m+1}{2}, \quad \frac{df(x)}{dx} > 0$$

can be obtained. So the value of $\rho(x)$ is monotone decreasing to the left of x^*. The value of $\rho(x)$ is monotone increasing to the right of x^*. This completes the proof of Theorem 5.3.

Suppose $x \to x^*$ in order to increase the consistency of the linguistic comparison matrix according to Theorem 5.3, and the consensus degree gradually rises when $x \to x^*$.

Theorem 5.4

Adjust the elements of the linguistic comparison matrix according to x^*'s direction; satisfactory consistency can be obtained.

Proof

The optimal solution x^* makes $\rho \to \min$, $\mu \to \max$ according to Formula (5.3). It is consistent after adjusting $\mu' \geq \mu$ in the linguistic comparison matrix. The adjustment ends when $\rho = 0$ according to Formula (5.3). So the adjustment method is convergent.

The method of this chapter points out the possible adjustment direction of some elements. Decision makers can make proper adjustments in mathematical meaning according to their own preferences. In certain situations, they can adjust element to x^* directly.

5.2.4   Group Aggregation Property of Linguistic Comparison Matrix

Consider a group decision-making problem with linguistic preference relations. Let $D = \{d_1, d_2, \ldots, d_m\}$ be the set of decision makers, and $\lambda_k, k = 1, \ldots, m$ be the weight vector of decision makers, where $\lambda_k \geq 0$, $\sum_{k=1}^{m} \lambda_k = 1$. Let $A_k, k = 1, \ldots, m$ be the linguistic preference relations provided by m decision makers $d_k, k = 1, \ldots, m$, where $A_k = (a_{ij}^{\ k})_{n \times n}$, $a_{ij}^{\ k} \in S$, then denote $\overline{A} = (\overline{a_{ij}})_{n \times n} = \lambda_1 A_1 + \cdots + \lambda_m A_m$ as the collective linguistic preference relation of $A_k, k = 1, \ldots, m$, where $\overline{a_{ij}} = \lambda_1 a_{ij}^{\ 1} + \cdots + \lambda_m a_{ij}^{\ m}$, $i, j = 1, \ldots, n$.

Theorem 5.5

Give the value of x_1, x_2, and suppose A_k's consistency meets $\mu(A_k) \leq \mu$, thus $\mu(\overline{A}) \leq \mu$.

Proof

$$\mu(\overline{A}) = \frac{\rho(\overline{A}) - x_2}{x_1 - x_2} = \frac{\rho(\overline{A})}{x_1 - x_2} - \frac{x_2}{x_1 - x_2},$$

then

$$\rho(\overline{A}) = \frac{\displaystyle\sum_{1 \leq i < j < k \leq n} \left| I(\overline{a_{ij}}) - I(\overline{a_{ik}}) + I(\overline{a_{jk}}) \right|}{C_n^3}$$

$$= \frac{\displaystyle\sum_{1 \leq i < j < k \leq n} \left| I(\sum_{t=1}^{m} \lambda_t a_{ij}^{\ t}) - I(\sum_{t=1}^{m} \lambda_t a_{ik}^{\ t}) + I(\sum_{t=1}^{m} \lambda_t a_{jk}^{\ t}) \right|}{C_n^3}$$

$$= \frac{\displaystyle\sum_{1 \leq i < j < k \leq n} \left| I(\lambda_1 a_{ij}^{\ 1}) - I(\lambda_1 a_{ik}^{\ 1}) + I(\lambda_1 a_{jk}^{\ 1}) + \cdots + I(\lambda_m a_{ij}^{\ m}) - I(\lambda_m a_{ik}^{\ m}) + I(\lambda_m a_{jk}^{\ m}) \right|}{C_n^3}$$

$$\leq \frac{\displaystyle\sum_{1 \leq i < j < k \leq n} \left| I(\lambda_1 a_{ij}^{\ 1}) - I(\lambda_1 a_{ik}^{\ 1}) + I(\lambda_1 a_{jk}^{\ 1}) \right| + \cdots + \left| I(\lambda_m a_{ij}^{\ m}) - I(\lambda_m a_{ik}^{\ m}) + I(\lambda_m a_{jk}^{\ m}) \right|}{C_n^3}$$

$$\Rightarrow \mu(\overline{A}) \leq \frac{\displaystyle\sum_{1 \leq i < j < k \leq n} \left| I(\lambda_1 a_{ij}^{\ 1}) - I(\lambda_1 a_{ik}^{\ 1}) + I(\lambda_1 a_{jk}^{\ 1}) \right| + \cdots + \left| I(\lambda_m a_{ij}^{\ m}) - I(\lambda_m a_{ik}^{\ m}) + I(\lambda_m a_{jk}^{\ m}) \right|}{(x_1 - x_2)C_n^3}$$

$$- \frac{x_2}{x_1 - x_2} \mu(\overline{A}) \leq \lambda_1 \mu + \cdots + \lambda_m \mu = \mu \sum_{k=1}^{m} \lambda_k = \mu$$

5.2.5   *Definition of Transitive Linguistic Preference and Adjustment*

The transitive linguistic preference relation is considered as a pre-requisite for the accuracy of decision making. In this chapter, an approach to testing whether the matrix is transitive consistent or not is discussed and an improved proposal is made for the inconsistent matrix.

Definition 5.4

Let $A = (a_{ij})_{n \times n}$ be a linguistic comparison matrix. A is called a *transitive consistent matrix* if there exists $a_{ij} > s_0$ and $a_{jk} > s_0$, then $a_{ik} > s_0$ for $i, j, k = 1, \ldots, n$.

Whether the linguistic comparison matrix is transitive consistent or not can be determined according to Definition 5.4. But it is difficult to determine directly when enormous numbers of alternatives are available. In the existing literature [85, 111], the order consistency test approach to the linguistic comparison matrix is studied based on graph theory. In this chapter, the linguistic comparison matrix is turned into a 0-1 preference matrix

$$R = (r_{ij})_{n \times n}, \begin{cases} a_{ij} > s_0 \rightarrow r_{ij} = 1 \\ a_{ij} < s_0 \rightarrow r_{ij} = 0. \\ a_{ij} = s_0 \rightarrow r_{ij} = 0 \end{cases}$$

Definition 5.5

The Boolean operator "sum" can be defined as $0 + 0 = 0$, $0 + 1 = 1$, $1 + 0 = 1$, $1 + 1 = 1$.

A conclusion can be obtained if the k time-exponentiation R^k of matrix R is calculated by Boolean operator; then $r_{ij}^k = 1$ means that there are no more than k side-directed chains between node i and node j; otherwise, $r_{ij}^k = 0$. In particular, $r_{ii}^k = 1$ means that the directed arcs start from node i and return to node i. In this case, it becomes a cycle chain. In this chapter, the number of the directed arcs is called the *side length*. For example, $i \rightarrow f \rightarrow l \rightarrow i$ is a cycle chain of three linked side lengths. In a directed graph, if there are a number of cycle chains and the number of the directed arcs is m,

m will be considered the influential number of order consistency of directed arc $i \rightarrow j$.

Theorem 5.6

For the directed graph of matrix R, if there is a cycle chain that has more than three linked side lengths, it can become a cycle chain that has three linked side lengths.

Proof

Let $i \rightarrow j \rightarrow f \rightarrow l \rightarrow i$ be assumed as four linked side lengths. For node i and node f, if there is $i \rightarrow f$, the cycle chain $i \rightarrow f \rightarrow l \rightarrow i$ can be set up whose side length is three. If there is $f \rightarrow i$, the cycle chain $i \rightarrow j \rightarrow f \rightarrow i$ can be set up. The proof process of the cycle chain that has four linked side lengths is similar.

Theorem 5.7

If there exists a 1 in the diagonal position of R^3, the corresponding linguistic comparison matrix is not transitive consistent. Otherwise, it is transitive consistent.

Proof

According to Theorem 5.6, if there is a cycle chain that has more than three linked side lengths, it can become a cycle chain that has three linked side lengths. The cycle chain that has three linked side lengths can be expressed as R^3. R^3 can be calculated to find the cycle chain from node i to node i in a directed graph. If there is $r_{ii}^3 = 1$ in R^3, the corresponding linguistic comparison matrix is not transitive consistent. If $r_{ii}^3 = 0$ for all i, the corresponding linguistic comparison matrix is transitive consistent.

The reach ability matrix T $(= R + \cdots + R^n)$ is proposed to test the order consistency. The calculation of R^3 can check the consistency based on Theorem 5.7 and the method of reachability matrix can deal only with the comparison matrix that does not have s_0 except diagonal positional elements. Otherwise, these methods are invalid. When s_0 is in a nondiagonal position, the method of finding cycle chain is used in this chapter. It can be shown as the following pseudo-code:

For $i, j, k = 1, i \neq j \neq k$ **TO** n

    If $(a_{ij} - s_0)(a_{ik} - s_0) \leq 0$ and $(a_{ik} - s_0)(a_{jk} - s_0) < 0$ then
        Print $i - j - k - i$;
    If $a_{ij} = s_0$, $a_{ik} = s_0$ and $a_{jk} \neq s_0$ then **Print** $i - j - k - i$;

End For

The following principle will be considered when the matrix elements that comprise the condition of cycle chain are revised.

1. Revise the directed arc prior when the affected number is greater.
2. Revise the judgment, which is close to s_0 prior when the affected number is equal.

Revising arcs of the most influential number can eliminate the number of cycle chains to the greatest degree. Generally, if the judgment approaches s_0, it means the fit and unfit quality degrees are equal for two elements. It is difficult for decision makers to decide and also leads to inconsistent judgment. In addition, in order to keep the linguistic comparison matrix's order consistency when adjusting the satisfaction consistency, the transitive consistency has to be checked. For example, $s_2 \to s_{-3}$. The comparison matrix's transitive consistence is the lowest condition, and some authors think that when the linguistic comparison matrix's consistency is checked, the transitive consistency has to be checked first.

5.2.6 Example Analysis

The satisfaction consistency improvements of Examples 5.1 and 5.2 do not considered the transitive consistency for convenience.

Example 5.1

This example is from Dong et al. [100], where

$$A = \begin{bmatrix} s_0 & s_2 & s_1 & s_1 \\ s_{-2} & s_0 & s_4 & s_2 \\ s_{-1} & s_{-4} & s_0 & s_4 \\ s_{-1} & s_{-2} & s_{-4} & s_0 \end{bmatrix}.$$

$\rho = 4.5$ can be obtained. Suppose $x_1 = 0.5, x_2 = 2$. Then the satisfaction consistency of A is -1.67. The consistency is poor.

The first calculation:

For

$$I(a_{12}) : \left| x - I(a_{13} + a_{32}) \right| + \left| x - I(a_{14} + a_{42}) \right| = \left| x + 3 \right| + \left| x + 1 \right|;$$

For

$$I(a_{13}) : \left| x - I(a_{12} + a_{23}) \right| + \left| x - I(a_{14} + a_{43}) \right| = \left| x - 6 \right| + \left| x + 3 \right|;$$

For

$$I(a_{14}) : \left| x - I(a_{12} + a_{24}) \right| + \left| x - I(a_{13} + a_{34}) \right| = \left| x - 4 \right| + \left| x - 5 \right|;$$

For

$$I(a_{23}) : \left| x - I(a_{21} + a_{13}) \right| + \left| x - I(a_{24} + a_{43}) \right| = \left| x + 1 \right| + \left| x + 2 \right|;$$

For

$$I(a_{24}) : \left| x - I(a_{21} + a_{14}) \right| + \left| x - I(a_{23} + a_{34}) \right| = \left| x + 1 \right| + \left| x - 8 \right|;$$

For

$$I(a_{34}) : \left| x - I(a_{31} + a_{14}) \right| + \left| x - I(a_{32} + a_{24}) \right| = \left| x \right| + \left| x + 2 \right|;$$

Table 5.1 expresses the revision program.

Suppose the adjustment is $a_{23} \rightarrow s_{-1}, a_{32} = s_1$, then $\rho = 2, \mu = 0$ can be obtained.

The second calculation:

For

$$I(a_{12}) : \left| x - I(a_{13} + a_{32}) \right| + \left| x - I(a_{14} + a_{42}) \right| = \left| x - 2 \right| + \left| x + 1 \right|;$$

For

$$I(a_{13}) : \left| x - I(a_{12} + a_{23}) \right| + \left| x - I(a_{14} + a_{43}) \right| = \left| x - 1 \right| + \left| x + 3 \right|;$$

For

$$I(a_{14}) : \left| x - I(a_{12} + a_{24}) \right| + \left| x - I(a_{13} + a_{34}) \right| = \left| x - 4 \right| + \left| x - 5 \right|;$$

Table 5.1 Adjusting the Elements of the Linguistic Comparison Matrix (First Calculation)

	$I(a_{12})$	$I(a_{13})$	$I(a_{14})$	$I(a_{23})$	$I(a_{24})$	$I(a_{34})$
x^*	$-3,-1$	$-3,6$	$4,5$	$-2,-1$	$-1,8$	$-2,0$
ρ	3	4.5	2.75	2	4.5	3
μ	-0.67	-1.67	-0.5	0	-1.67	-0.67

Table 5.2 Adjusting the Elements of the Linguistic Comparison Matrix (Second Calculation)

	$I(a_{12})$	$I(a_{13})$	$I(a_{14})$	$I(a_{23})$	$I(a_{24})$	$I(a_{34})$
x^*	$-1,2$	$-3,1$	$4,5$	$-2,-1$	$-1,3$	$0,3$
ρ	2	2	0.5	2	2	1.5
μ	0	0	1	0	0	0.33

For

$$I(a_{23}): |x - I(a_{21} + a_{13})| + |x - I(a_{24} + a_{43})| = |x+1| + |x+2|;$$

For

$$I(a_{24}): |x - I(a_{21} + a_{14})| + |x - I(a_{23} + a_{34})| = |x+1| + |x-3|;$$

For

$$I(a_{34}): |x - I(a_{31} + a_{14})| + |x - I(a_{32} + a_{24})| = |x| + |x-3|;$$

Table 5.2 expresses the revision program.

Suppose $a_{14} \rightarrow s_4, a_{41} = s_{-4}$ is adjusted, then $\rho = 0.5$, $\mu = 1$ can be obtained. So the satisfaction consistency of the linguistic comparison matrix is 1 and these two adjustments have involved two pairs of elements.

In Dong et al. [100], the consequences of the two improvements are

$$\begin{bmatrix} s_0 & s_{1.4} & s_1 & s_{1.6} \\ s_{-1.4} & s_0 & s_3 & s_{2.4} \\ s_{-1} & s_{-3} & s_0 & s_3 \\ s_{-1.6} & s_{-2.4} & s_{-3} & s_0 \end{bmatrix}$$

and

$$\begin{bmatrix} s_0 & s_2 & s_1 & s_3 \\ s_{-2} & s_0 & s_3 & s_2 \\ s_{-1} & s_{-3} & s_0 & s_3 \\ s_{-2} & s_{-2} & s_{-3} & s_0 \end{bmatrix}$$

Moreover, $\rho = 2.9$ and $\rho = 2.5$ can be obtained. The consistent degrees are -0.6 and -0.33. The consequences of Dong et al. [100] are still poor based on the satisfaction consistency of this chapter. In addition, most of the elements that are judged by experts are reliable. Therefore, adjustment of the comparison matrix should involve only a few elements. But the adjusted consequences of the first kind of optimization model adjust all elements except the diagonal elements in Dong et al. [100], so the scope of the adjustment is too large.

Example 5.2

The following comparison matrix is another example in Dong et al. [100].

$$B = \begin{bmatrix} S_0 & S_0 & S_2 & S_{-1} & S_4 \\ S_0 & S_0 & S_{-1} & S_0 & S_3 \\ S_{-2} & S_1 & S_0 & S_2 & S_1 \\ S_1 & S_0 & S_{-2} & S_0 & S_2 \\ S_{-4} & S_{-3} & S_{-1} & S_{-2} & S_0 \end{bmatrix} \quad A_1 = \begin{bmatrix} S_0 & S_{-1} & S_3 & S_{-1} & S_3 \\ S_1 & S_0 & S_1 & S_0 & S_2 \\ S_{-3} & S_{-1} & S_0 & S_{-1} & S_2 \\ S_1 & S_0 & S_1 & S_0 & S_0 \\ S_{-3} & S_{-2} & S_{-2} & S_0 & S_0 \end{bmatrix}$$

$$A_2 = \begin{bmatrix} S_0 & S_1 & S_2 & S_0 & S_4 \\ S_{-1} & S_0 & S_{-1} & S_0 & S_0 \\ S_{-2} & S_1 & S_0 & S_{-1} & S_3 \\ S_0 & S_0 & S_1 & S_0 & S_1 \\ S_{-4} & S_0 & S_{-3} & S_{-1} & S_0 \end{bmatrix} \quad A_3 = \begin{bmatrix} S_0 & S_0 & S_3 & S_1 & S_3 \\ S_0 & S_0 & S_{-2} & S_2 & S_2 \\ S_{-3} & S_2 & S_0 & S_1 & S_1 \\ S_{-1} & S_{-2} & S_{-1} & S_0 & S_{-1} \\ S_{-3} & S_{-2} & S_{-1} & S_1 & S_0 \end{bmatrix}$$

$$A_4 = \begin{bmatrix} S_0 & S_2 & S_0 & S_{-1} & S_2 \\ S_{-2} & S_0 & S_{-1} & S_1 & S_0 \\ S_0 & S_1 & S_0 & S_{-1} & S_2 \\ S_1 & S_{-1} & S_1 & S_0 & S_1 \\ S_{-2} & S_0 & S_{-2} & S_{-1} & S_0 \end{bmatrix}$$

Suppose $x_1 = 0.5$, $x_2 = 2$ and the consistency index can be showed as Table 5.3.

Table 5.3 The Consistency of the Linguistic Comparison Matrix

		B	A_1	A_2	A_3	A_4	$\bar{A}$	$\bar{\bar{A}}$
1	1-2-3	3	3	2	5	1	2.25	2.7
2	1-2-4	1	0	1	1	0	1.5	0.9
3	1-2-5	2	2	3	1	0	1.5	1.9
4	1-3-4	5	3	1	3	0	1.75	2.1
5	1-3-5	1	1	1	1	0	1	1.3
6	1-4-5	3	4	3	3	2	3	3.3
7	2-3-4	1	0	2	3	3	2	1.5
8	2-3-5	3	1	2	3	1	0.25	0.5
9	2-4-5	1	2	1	1	2	0	0.5
10	3-4-5	3	3	3	1	2	2.25	2.5
	ρ	2.3	1.9	2.1	2.2	1.1	1.55	1.72
	μ	−0.2	0.067	−0.067	−0.133	0.6	0.3	0.187

The satisfaction consistencies of $\bar{A}$ and $\bar{\bar{A}}$ are based on the satisfaction consistency of this chapter. They are 0.3 and 0.7. In addition, $\bar{A}$ and $\bar{\bar{A}}$ are not transitive consistent. For $\bar{A}$, there are 1–2–4: $a_{12} = s_{0.5}$, $a_{14} = s_{-0.25}$, $a_{24} = s_{0.75}$ and 2–3–4: $a_{23} = s_{-0.75}$, $a_{24} = s_{0.75}$, $a_{34} = s_{-0.5}$. For $\bar{\bar{A}}$, there are 1– 2– 4: $a_{12} = s_{0.1}$, $a_{14} = s_{-0.3}$, $a_{24} = s_{0.5}$; 2–3–4: $a_{23} = s_{-0.4}$, $a_{24} = s_{0.5}$, $a_{34} = s_{-0.6}$.

Example 5.3

A venture capital company has a fund to carry on the most superior investment. There are five alternatives: a biopharmaceutical company, a food company, a fashion company, a computer software company, and a real estate investment company. The multiple comparative linguistic comparison matrix of each option is given by decision makers as follows:

$$F = \begin{bmatrix} s_0 & s_2 & s_{-2} & s_0 & s_3 \\ s_{-2} & s_0 & s_{-1} & s_0 & s_4 \\ s_2 & s_1 & s_0 & s_{-1} & s_6 \\ s_0 & s_0 & s_1 & s_0 & s_2 \\ s_{-3} & s_{-4} & s_{-6} & s_{-2} & s_0 \end{bmatrix}$$

Is the linguistic comparison information given by decision makers based on logical consistency? An inspection is implemented as follows:

Step 1: Check the order consistency. There is s_0 in a nondiagonal position of linguistic comparison matrix F. So use the method of finding cycle chains to determine whether it is transitive consistent or not. The number of cycle chains of the comparison matrix is three based on calculation. As a result, this linguistic comparison matrix is not transitive consistent.

Step 2: Improve the order consistency. Find all cycle chains in the linguistic comparison matrix. They are 1–2–4–1, 1–3–4–1, and 2–3–4–2. The influential number of 2–4, 4–1, 3–4 is two. 2–4, 1–4 are the judgments of s_0. This means that these two alternatives are equal. These two judgments have to be adjusted in advance based on the principle of order consistency improvement. Suppose the decision maker changes $a_{24} = s_0$ to s_{-1}, then two cycle chains are deleted.

Repeat step 1 and find the cycle chain. The comparison matrix is still not transitive consistent. The chain 1–3–4–1 is found such that $a_{13} = s_{-2}$, $a_{14} = s_0$, $a_{34} = s_{-1}$. a_{14} is the judgment of s_0. The results must be returned to the decision maker to check. Suppose the decision maker changes $a_{14} = s_0$ to s_{-1}, then all cycle chains are deleted. Then

$$F = \begin{bmatrix} s_0 & s_2 & s_{-2} & s_{-1} & s_3 \\ s_{-2} & s_0 & s_{-1} & s_{-1} & s_4 \\ s_2 & s_1 & s_0 & s_{-1} & s_6 \\ s_1 & s_1 & s_1 & s_0 & s_2 \\ s_{-3} & s_{-4} & s_{-6} & s_{-2} & s_0 \end{bmatrix}$$

can be obtained.

Table 5.4 Optimal Improved Values

	a_{12}	a_{13}	a_{14}	a_{15}	a_{23}	a_{24}	a_{25}	a_{34}	a_{35}	a_{45}
x^*	−1	0	1	4	−2	−2	1	0	5	5
ρ	1.6	2.1	2.1	2.2	2.2	2.2	2	2	2	1.6
μ	0.267	−0.067	−0.067	−0.133	−0.133	−0.133	0	0	0	0.267

Step 3: Check the satisfaction consistency. Suppose $x_1 = 0.5$, $x_2 = 2$. Check the satisfaction consistency of the linguistic comparison matrix after improving its order consistency. Then $\rho = 2.3$ can be obtained. The consistency degree is −0.2. This means that the consistency is poor and needs modification.

Step 4: Improve the satisfaction consistency. Calculate x^* to obtain the consequence in Table 5.4.

In Table 5.4, if $a_{12} = s_2 \rightarrow s_{-1}$ or $a_{45} = s_2 \rightarrow s_5$, the consistency degree can be improved to 0.267. Suppose the decision makers modify $a_{45} = s_2 \rightarrow s_5$, and the consistency degree is still poor. Repeat this process again. Modifying $a_{12} = s_2 \rightarrow s_{-1}$, the consistency degree is improved to 0.733. If the modification of a_{12} crosses s_0 (i.e., the modification from s_2 to s_{-1} and s_0 is less than s_2 and more than s_{-1}) the order consistency needs to be checked again. There is no s_0 in a nondiagonal position of the linguistic comparison matrix. So calculate R^3, and the comparison matrix is transitive consistent after modifying

$$F = \begin{bmatrix} s_0 & s_{-1} & s_{-2} & s_{-1} & s_3 \\ s_1 & s_0 & s_{-1} & s_{-1} & s_4 \\ s_2 & s_1 & s_0 & s_{-1} & s_6 \\ s_1 & s_1 & s_1 & s_0 & s_5 \\ s_{-3} & s_{-4} & s_{-6} & s_{-5} & s_0 \end{bmatrix}.$$

This satisfies the transitive consistency and satisfaction consistency.

5.3 Aggregation Method of Linguistic Evaluation on Multiple-Criteria Decision Making Based on Duality Semantic Transform

5.3.1 Background and Problems

In many of the actual processes of decision making, the property of suggestions is usually expressed in quantitative (real number, interval number, triangular fuzzy number) or qualitative ways. Because of the fuzziness and uncertainty of alternatives, decision makers usually use linguistic evaluation to express preference. For example, in the integrative evaluation for the leader of management quality, it is difficult for decision makers to express their understanding of properties in quantitative numbers, so they use "very poor," "good," "very good" to evaluate. In recent years, group decision based on linguistic evaluation received the attention of scholars. The biggest problem in deciding linguistic evaluation is how to translate the words' information into easy decision-making ways effectively. Usually, triangular fuzzy number, interval number, trapezium fuzzy number, or the linguistic model is adopted to deal with linguistic preference.

The first method requires assuming a function under the operation of fuzzy number. This method usually has an indirect relation of former linguistic evaluation; it will increase the fuzziness of translation and make former linguistic evaluation distortion.

The second method assumes that language assessment is discrete; it makes the original language information lose the original meaning of information. Therefore, Dr. Herrera [55, 67] proposed a two-element semantic method about collecting language information. He defined the two-element semantic, which overcame the disadvantage of former research. In recent years, two-element semantic group decisions based on language information put particular emphasis on processing two-element semantic with weights of quality and experts given. In fact, in the complex decision-making process, decision makers and property weight is a rather difficult question. Whether property weight and decision makers' weights are equal will threaten the authenticity of decision.

This section first puts the language of information of the decision maker into the corresponding semantic judge matrix. Then, according to the matrix expressed preferences of decision makers, it uses a planning model method of the two-element semantic to solve property weight. Solving decision makers' weight is done by gathering and comparing the executives' tastes and the overall preferences. Finally, the projects are gathered to order, and the group decision method of solving the linguistic information is given.

This section defines the degree of consensus of decision makers to measure the degree of consistency of decision makers. It requires the decision maker who has declination of degree to improve his preference to make the group decision coincident. Finally, the method is used in foreign trade risk management assessment program choice incident.

5.3.2 Duality Semantic of the Linguistic Evaluation Preference

In considering of the target group of decision-making problems, suppose a limited scheme set $X = \{x_1, x_2, \ldots, x_m\}$, where x_i is means the i decision proposal; property set $A = \{a_1, a_2, \ldots, a_n\}$, where a_j is means the j property indicator; $P = \{p^1, p^2, \ldots, p^k\}$ is the groups decision set, where p^k is the k decision maker. Decision makers use linguistic evaluation set S to express the preference degree of property indicator. This means the decision maker k chooses one element in linguistic evaluation set S to express the preference degree of j in project i, recorded as p_{ij}^k. The reference of linguistic evaluation set S in this section has seven languages of set phrases. It is

$$S = \begin{cases} s_0 = N\left(none\right), s_1 = VL\left(very \quad low\right), s_2 = L\left(low\right), s_3 \\ = M\left(medium\right), s_4 = H\left(high\right), \\ s_5 = VH\left(very \; high\right), s_6 = P\left(perfect\right) \end{cases}$$

The linguistic set has certain characteristics:

Ordered nature, when $i \geq j$, then $S_i \geq S_j$.

Has inverse operation operator Neg, when $j = T - i$, then $Neg\left(S_i\right) = S_j$, $T + 1$ means the number of elements in set S.

Operation of maximation and minimization, when $S_i \geq S_j$, then $Max\left\{S_i, S_j\right\} = S_i$; $Min\left\{S_i, S_j\right\} = S_j$.

It also adjusts to the following definition.

Definition 5.6

If $S_k \in S$ is a description of linguistic information, its semantic form can be obtained by function θ

$$\theta : S \rightarrow S \times \left[-0.5, 0.5\right] \tag{5.7}$$

$$\theta\left(s_i\right) = \left(s_i, 0\right), s_i \in S \tag{5.8}$$

Definition 5.7

Suppose a real number $\beta \in [0,T]$ is obtained by some gathering method of linguistic evaluation set S, and $T+1$ is the number of elements in linguistic evaluation set S, then β expresses semantic information by function Δ.

$$\Delta : [0,T] \rightarrow S \times [-0.5, 0.5] \qquad (5.9)$$

$$\Delta(\beta) = \begin{cases} s_k, k = \text{round}(\beta) \\ \alpha_k = \beta - k, \alpha_k \in [-0.5, 0.5] \end{cases} \qquad (5.10)$$

where *round* is the rounding operator; α_k is the symbolic translation. It means the deviation of results of evaluations with s_k.

Definition 5.8

Suppose (s_k, α_k) is the semantic, s_k is the k element in S, $\alpha_k \in [-0.5, 0.5]$, then it has one inverse function Δ^{-1} to make it into the number β.

$$\Delta^{-1} : S \times [-0.5, 0.5] \rightarrow [0,T] \qquad (5.11)$$

$$\Delta^{-1}(s_k, \alpha_k) = k + \alpha_k = \beta \qquad (5.12)$$

Suppose $(s_\alpha, \alpha_\alpha)$ and (s_β, α_β) are two semantics, then it has the following rules of the comparison of semantics:

If $\alpha < \beta$, then $(s_\alpha, \alpha_\alpha) < (s_\beta, \alpha_\beta)$;

If $\alpha = \beta$, $\alpha_\alpha = \alpha_\beta$, then $(s_\alpha, \alpha_\alpha) = (s_\beta, \alpha_\beta)$; $\alpha_\alpha < \alpha_\beta$, $(s_\alpha, \alpha_\alpha) < (s_\beta, \alpha_\beta)$; $\alpha_\alpha > \alpha_\beta$, $(s_\alpha, \alpha_\alpha) > (s_\beta, \alpha_\beta)$.

Definition 5.9

Suppose $(s_\alpha, \alpha_\alpha)$ and (s_β, α_β) are two semantics at random, so the distance of them is

$$d\left[(s_\alpha, \alpha_\alpha), (s_\beta, \alpha_\beta)\right] = \Delta\left\{\left[\Delta^{-1}(s_\alpha, \alpha_\alpha) - \Delta^{-1}(s_\beta, \alpha_\beta)\right]^2\right\} \qquad (5.13)$$

Definition 5.10

(Judge of degree of decision maker): Suppose it has $k(k = 1, 2, \ldots, K)$ decision makers evaluating $X = \{x_1, x_2, \ldots, x_m\}$ projects, define

$$
\varepsilon^k = \left. \left(r - \sum_{i=1}^{m} \left| z_i(w) - z_i(w)^k \right| \right) \middle/ r \right. \tag{5.14}
$$

is the k decision maker's deviation with group makers. It is called the judge of degree of the k decision maker.

r Means marked degree of linguistic evaluation S; for example, linguistic evaluation set S has seven languages of set phrases (discussed above), then $r = 6$. $z_i(w)$ means the comprehensive index value of projects, and $z_i(w)^k$ means the comprehensive index value of the k decision maker. And

$$
z_i(w) = \Delta\left[\sum_{k=1}^{K} \lambda_k \sum_{j=1}^{n} w_j \Delta^{-1}\left(r_{ij}^k, 0\right) \right] i = 1, 2, \ldots, m, \lambda_k
$$

is the weight of the decision maker;

$$
z_i(w)^k = \Delta\left[\sum_{j=1}^{n} w_j^k \Delta^{-1}\left(r_{ij}^k, 0\right) \right] i = 1, 2, \ldots, m.
$$

It is obtained that if one decision maker has no deviation with groups, $\varepsilon^k = 1$, and if the deviation of one decision maker with groups is r, $\varepsilon^k = 0$. According to the definition, it is better to reflect the consensus degrees of the decision maker with groups, find the consensus of the inferior, and make him modify preferences information. Define $\varepsilon^k \geq 0.6$ when decision makers find consensus.

Definition 5.11

(Modification of deviation of decision makers): Regarding to the decision makers' preferences that did not meet the consensus, define

$$
\left| z_i(w) - z_i(w)^k \right| \geq a
$$

when decision makers need to modify their judgment to reach the agreement with groups. a is the threshold of deviation of one decision maker comprehensive index value and groups. When $a \geq 1$, make the decision maker modify his preference to reach agreement with groups. $a \geq 1$ Means that the deviation of the decision maker with groups is more than one marked degree; it means there is large deviation between the preferences of the decision maker and group in one project. We can suppose a different threshold in a different situation: if the precision required higher in decision, a should be smaller; if the precision required lower in decision, a should be larger.

Definition 5.12

Suppose it has $k(k = 1, 2, \dots, K)$ decision makers evaluate $X = \{x_1, x_2, \dots, x_m\}$ projects,

$$\varepsilon = \frac{\sum_{k=1}^{K} \varepsilon^k}{K}$$

is called the average degree of group. Generally, when $\varepsilon \geq 0.6$, it is thought that the group reach agreement.

5.3.3 Aggregation Method Based on Duality Semantic Transform

According to the properties of semantic and the definition of distance, a method is given about linguistic evaluation in the group decision problem. The key of this method is solving the weights of property and decision makers. According to the preference expressed by the judgment matrix of every decision maker, a model is built to determine the properties of semantic and the property weights of every decision maker. Then the weights of decision makers are calculated by gathering and comparing preferences of all decision makers and groups. The steps are as follows:

Step 1: **Linguistic evaluation information is changed into a binary.** Make the linguistic evaluation information of decision makers into the semantic form. Make the evaluation information of the k decision maker p_{ij}^k into $\left(r_{ij}^k, \alpha\right)$. If $R_{ij}^k \in S$, then $\alpha = 0$; if $R_{ij}^k \notin S$, then $\left(r_{ij}^k, \alpha\right)$. Then express the linguistic evaluation of all decision makers by semantic matrix form.

Step 2: The first solving of property weights for each decision maker. Assuming each decision makers' weight is equal, it means each weight of k decision makers is

$$\frac{1}{k}.$$

Use the minimum preference of language information of policy makers as the principle to solve model $P_{5.1}$ and solve property weights with the decision maker weights being equal. This section solves problems of plus–minus deviation of distance formula; use the square method to solve this problem.

The property weights model with equal decision maker weights is $P_{5.1}$:

$$\min \sum_{j=1}^{n}\sum_{i=1}^{m}\sum_{\substack{k=1\,p>k}}^{K} \Delta\left[\lambda_k\Delta^{-1}\left(r_{ij}^k,0\right)w_j^k - \lambda_p\Delta^{-1}\left(r_{ij}^p,0\right)w_j^p\right]^2$$

$$s.t. \quad \sum_{j=1}^{n} w_j^k = 1,\ k = 1,\ 2,\ \dots,\ K$$

$$w_j^k \geq 0$$

$$i = 1,\ 2,\ \dots,\ m;\ j = 1,\ 2,\ \dots,\ n;\ k = 1,\ 2,\ \dots,\ K$$

w_j^k means the weight of the k decision maker to the j property; w_j^p means the weight of the p decision maker to the j property; λ_k means the weight of the k decision maker; λ_p means the weight of the p decision maker. First, suppose the weights of decision makers are equal; it means

$$\lambda_k = \lambda_p = \frac{1}{K}.$$

In $P_{5.1}$, the object function means the sum of preference of comprehensive index value is minimum, then solve the property weights of makers.

Theorem 5.8

In each type of calculation for property weights, preference of language information of policy makers can get the minimum, so it can get the optimization solution.

Step 3: The trial of decision makers to solve the weights.
Concentrate the preference of every decision maker by Formula (5.15); finally, get the sort of scheme of each decision maker.

$$z_i\left(w\right) = \Delta\left[\sum_{j=1}^{n} w_{ij}^{k}\Delta^{-1}\left[r_{ij}^{k},0\right]\right] i = 1, 2, \ldots, m \qquad (5.15)$$

Concentrate the preference of every decision maker and solve the inverse of preferences valid values by semantic inverse operation. According to the rule of semantic comparison in Definition 5.8, get the sort of scheme of each decision maker.

Step 4: Assessment of group to solve attributes' weights. First suppose decision maker weights are equal,

$$\lambda_k = \lambda_p = \frac{1}{K}.$$

Solve comprehensive weights by model $P_{5.2}$.

$$\min \sum_{j=1}^{n}\sum_{i=1}^{m}\sum_{k=1\,p>k}^{K} \Delta\left[\lambda_k\Delta^{-1}\left(r_{ij}^{k},0\right)w_j - \lambda_p\Delta^{-1}\left(r_{ij}^{p},0\right)w_j\right]^2$$

$$s.t. \quad \sum_{j=1}^{n} w_j = 1, \, j = 1, 2, \ldots, n$$

$$w_j \geq 0$$

$$i = 1, 2, \ldots, m; \, j = 1, 2, \ldots, n; \, k = 1, 2, \ldots, K$$

w_j means the j property weight. k, p and the meaning of objective function are similar to model $P_{5.1}$.

Step 5: The trial of decision makers to solve the comprehensive index value. Solve comprehensive property weights by model $P_{5.2}$, first concentrating the comprehensive value of projects with decision-maker weights that are equal.

$$z_i\left(w\right)^k = \Delta\left[\sum_{j=1}^{n} w_j^k \Delta^{-1}\left(r_{ij}^k, 0\right)\right] i = 1, 2, \dots, m \qquad (5.16)$$

Calculate judge of degrees of all the decision makers. Test whether the deviation of the policy maker decision is too large for the group to determine. If there is a decision maker with a large deviation, adjust the decision of the decision maker, then calculate the average degree of consensus of groups. The importance of this section is pointing out the decision maker with inconsistent value in order to get the same degree quickly.

Step 6: Estimation of decision makers' weight. Compare the comprehensive index value of each project in step 3 and that of groups in step 5. According to the rule of minimum deviation of the comprehensive index value after introducing decision maker weights and group index value, the model for calculating decision weight λ_k can be found:

$$\min \sum_{i=1}^{m}\left[\sum_{k=1}^{K} z_i^k\left(w\right)\lambda_k - z_i\left(w\right)\right]^2$$

$$s.t. \sum_{k=1}^{K} \lambda_k = 1 \qquad\qquad P_{5.3}$$

$$\lambda_k \geq 0.05; k = 1, 2, \dots, K$$

$\sum_{k=1}^{K} z_i^k\left(w\right)\lambda_k$ is the comprehensive index value of decision maker with weight, $z_i\left(w\right)$ is the comprehensive index value of group to project, and $\lambda_k \geq 0.05$ can ensure that each decision maker joins this decision.

Theorem 5.9

In the process of calculating decision maker weights, the deviation of the comprehensive index value after introducing decision maker weights and group index value can get the minimum, which means there is minimum in model $P_{5.3}$.

Step 7: Ranking of the alternatives. Put the weight of every decision maker of model $P_{5.3}$ into model $P_{5.2}$, calculate the property weight of groups again, and collect the comprehensive index value of every alternative. Last, calculate the rank of preference by the result of the comprehensive index value of each alternative. The alternative is better when the result of the comprehensive index value is larger and vice versa.

5.3.4   Example Analysis

With the development of a global economy, in areas with dense population and industries, the production, storage, transportation, and usage of inflammable, explosive, and poisonous chemical hazards are increasing. The risk environmental pollution accidents is also increasing; major accidents frequently occur. Hazardous-chemical accidents have the characteristic of abruptness, association of promotion of the diversity, severity of harm, difficulty of dealing with, and so on. Cities have their own development processes, and the ability of cities to deal with hazards problems is decreasing as populations increase. In 2008, there was an accident in Zhejiang province involving transport vehicles in a pileup with hazardous materials—a toxic chemical from a leaked strobe inside a vehicle. Emergency crews posed four alternatives for decreasing accidents involving dangerous chemicals: (x_1) isolate contaminated areas; collect chemicals into dry, clean, and covered containers using a clean shovel; (x_2) isolate contaminated areas, spray low-pressure water to dilute; (x_3) build a causeway or tunnel to drain the leaked chemicals; (x_4) use sandy soil to cover the chemicals to prevent pollution from invading the city. Survey the accident situation and spread of chemicals of the scene: (a_1), restore the surrounding environment and safety traffic as soon as possible (a_2), the scheme at the scene of rescue (a_3), the cost of the solution of rescue (a_4). Use different alternatives for different accidents. In this accident, five experts provided four emergency plans and four linguistic values to describe the degrees of satisfaction in meeting the objectives.

Step 1: Express the linguistic evaluations of five experts by the semantic matrix of evaluation.

$$
P^1: \begin{bmatrix} (H,0) & (M,0) & (H,0) & (VL,0) \\ (VH,0) & (N,0) & (P,0) & (M,0) \\ (L,0) & (VL,0) & (L,0) & (N,0) \\ (M,0) & (P,0) & (VH,0) & (VL,0) \end{bmatrix}
$$

$$
P^2: \begin{bmatrix} (M,0) & (VH,0) & (P,0) & (VL,0) \\ (H,0) & (M,0) & (N,0) & (M,0) \\ (L,0) & (VH,0) & (H,0) & (M,0) \\ (VL,0) & (P,0) & (VL,0) & (VH,0) \end{bmatrix}
$$

$$
P^3: \begin{bmatrix} (L,0) & (VH,0) & (H,0) & (M,0) \\ (VH,0) & (N,0) & (L,0) & (M,0) \\ (H,0) & (P,0) & (VL,0) & (H,0) \\ (M,0) & (L,0) & (VH,0) & (N,0) \end{bmatrix}
$$

$$
P^4: \begin{bmatrix} (H,0) & (VL,0) & (M,0) & (L,0) \\ (VH,0) & (M,0) & (VL,0) & (L,0) \\ (VL,0) & (VH,0) & (N,0) & (M,0) \\ (L,0) & (VL,0) & (M,0) & (N,0) \end{bmatrix}
$$

$$
P^5: \begin{bmatrix} (P,0) & (M,0) & (VL,0) & (H,0) \\ (VH,0) & (VL,0) & (N,0) & (M,0) \\ (L,0) & (M,0) & (M,0) & (L,0) \\ (VL,0) & (M,0) & (L,0) & (M,0) \end{bmatrix}
$$

Table 5.5 Weights of Criteria When Decision Maker Weights Are Equal

	W_{I1}	W_{I2}	W_{I3}	W_{I4}
P^1	0.448	0.122	0.085	0.345
P^2	0.630	0.098	0.111	0.161
P^3	0.456	0.136	0.167	0.241
P^4	0.468	0.116	0.176	0.240
P^5	0.374	0.149	0.149	0.150

Step 2: Solve property weights of each decision maker with decision maker weights equal. Detailed weights can be seen in Table 5.5.

This shows the differences of decision maker on property. Many decision makers think (a_1) is the most important, followed by (a_4), then (a_2) and (a_3). The characteristics of chemicals, the control of scene, and the tendency of future development are most important; the cost of emergency rescue is more important.

Step 3: Obtain the rank of emergency alternatives of decision makers in order to get the preference of emergency alternatives of decision makers.

$$P^1: x_2 > x_4 > x_1 > x_3;$$

$$P^2: x_2 > x_1 > x_3 > x_4;$$

$$P^3: x_3 > x_2 > x_1 > x_4;$$

$$P^4: x_2 > x_1 > x_3 > x_4;$$

$$P^5: x_1 > x_2 > x_3 > x_4.$$

Step 4: Obtain the property weights by model $P_{5.2}$.

$w_1 = 0.465$, $w_2 = 0.152$, $w_3 = 0.141$, $w_4 = 0.242$. By calculating the property weights, we can know that when experts assess hazardous material incidents, judgment is given by situation of scene and timeliness of project, then the convenience and getting the environment of the scene back to normal are considered.

Step 5: Calculate the rank of emergency alternatives by the property weights. This is the rank that is after giving comprehensive property $(M, 0.324), (M, 0.376), (L, 0.494), (L, 0.364)$, then $x_2 > x_1 > x_3 > x_4$.

Step 6: Calculate the decision maker weights by model $P_{5.3}$. $\lambda_1 = 0.216$, $\lambda_2 = 0.619$, $\lambda_3 = 0.065$, $\lambda_4 = 0.050$, $\lambda_5 = 0.050$. The second decision maker and the third decision maker have better preference degree with the group, so their weights are larger, whereas the fourth and fifth decision makers are not, so their weights are small.

Step 7: According to Definition 5.10, calculate the judge degree of decision makers. This can result in $\varepsilon^1 = 0.554$, $\varepsilon^2 = 0.898$, $\varepsilon^3 = 0.705$, $\varepsilon^4 = 0.688$, $\varepsilon^5 = 0.605$. The deviation of the first decision maker is larger because $\varepsilon^1 < 0.6$. The deviations of the first decision maker are $0.481, 0.409, 1.306, 0.482$, the first decision maker should adjust the judgment of (x_3). Suppose the first decision maker adjusts the first and second judgment of property of (x_3). The deviation translates from 1.306 to 0.6478; ε^1 translates from 0.554 to 0.663. The consensus on the average is 0.711, and the group consensus improves. After that, with the preference of the decision maker changed, calculate again and get the rank: $(M, 0.111), (M, 0.351), (M, -0.398), (L, 0.254)$, $x_2 > x_1 > x_3 > x_4$. In this accident, emergency alternative (x_2) should be the first alternatives; the next are emergency alternatives (x_1) and (x_3); emergency alternative (x_4) is the last one. In this accident, considering the kinds of chemicals, diffusion of chemicals, and timeliness of rescue, emergency alternative (x_2) should be the first alternative, and emergency alternative (x_1) is the backup alternative. This proves that the method posed in this chapter is in accordance with the alternatives of experts.

5.4   Summary and Future Research

That the comparison matrix is logical consistent is a prerequisite for scientific decision. The consistency problem of the linguistic comparison matrix is studied in this section. Although there are some outcomes of transitive consistency, the existing methods could not deal with the comparison matrix that has equal alternatives. The consistency degree index is proposed in the satisfactory of the linguistic comparison matrix. The optimal element values of the linguistic comparison matrix whose elements are not changed can be obtained. The definition of satisfaction consistency is simple and flexible in this chapter. The method for improving the satisfaction consistency is easy to understand. Compared with the existing literature, the improved direction is given in the process of consistency improvement and is easy to calculate. In addition, the improvement of transitive consistency is studied to provide suggestions for decision makers.

In order to solve the problem of linguistic evaluation information quickly, this chapter gives a method of gathering property and decision maker weights based on semantic information. This applies to less than ten decision makers, five to seven alternatives, and five to seven properties of group decision (consider restrictions of the effective judge matrix export preferences and solving model). On the one hand, the method of this chapter avoids the problem that emerged by language information processing before; on the other hand, we can get property and decision maker weights by model. It avoids the disadvantage of subjective weight before and gets the demand of solving emergency accidents.

This chapter pays attention to two aspects of linguistic preference: one is the consistency of the language judgment matrix; the other is the decision method of the language assessment. In fact, it can be seen from these studies that language preference is the hot point of theory study in decision, but there are many problems in linguistic evaluation. The representative aspects are following:

1. The consistency of language information: There are articles studying this, but their methods are not mature [85, 111], Language form has more intention than number form literally. If we study from number only, the language form will lose its importance.
2. The complex process of decision making can be simplified to three processes. First of all, decision makers think with the actual problem and express by language, then solve the

language information. It is the most effective way to reflect the thinking and behavioral characteristics of decision makers. In many cases, it is difficult for simple language to express the complex process of thinking, so it is necessary to improve the linguistic evaluation set.

3. The worst problem is that most articles translate the linguistic information into number form. The problem of language information actually is number form. This translation is a simple and effective method but not perfect. The study should start from the characteristics of language preference and study the decision method based on language preference and keeping the characteristics of language preference when translating language form to number form. Measuring the complex process of decision making by language form is a significant simplification, and the information has been translated. The next importance is that solving the language information that is original and the behavior of decision makers.

6

Aggregation Method on Multiple Style Preference for Group Decision Making

6.1 Aggregation Method of Hybrid Uncertain Comparison Matrix

6.1.1 Background

With the development of society, the progress of science and technology, and increase in information and knowledge, decision-making problems are becoming more complicated and more people are involved in the decision-making process. In order to reduce the workload of the group because many evaluation indexes and alternatives exist, the random judgment division can be used and integrated in general. A completely pairwise comparison matrix can be obtained based on the evaluation indices. For example, there are six evaluation alternatives $B_j, j = 1, \ldots, 6$ and six evaluation indices $F_i, i = 1, \ldots, 6$ in one decision-making problem. If the completely pairwise comparison is adopted to obtain the order of alternatives, the number of judgments will be $180(6 \times 5 \times 6 = 180)$. It is tiresome for the expert to sustain the judgment work. Furthermore, the judgment accuracy may fall after a number of pairwise comparisons. In the context of complex decision making, more decision-making alternatives and evaluation indices put much pressure on experts' judgment tasks. In this case, if several experts judge cooperatively for the decision-making problems, the amount of judgment work for each expert will be reduced to 60%. The lighter workload may make the judgments more detailed and accurate. The random division of the collaboration is shown as follows. First, the division of judgment tasks, for example, the alternatives B_1 and B_3, B_3 and B_5 based on the index F_1 can be assigned to expert 1; the alternatives B_2 and B_4, B_4 and B_5 based on the index F_1 can be assigned to expert 2. Then, expert 3 completes the rest of the judgment. To a great extent, this cooperative division reduces the judgment workload and the errors caused by experts' wrong individual judgments. In addition, this cooperative judgment style has the feature of checking the consistency of evaluation criteria and urges the experts to use the same judgment criteria. If a big difference is raised among the experts' understanding, they can coordinate judgments to obtain consistent judgments.

The cooperative division judgment style has another problem in that since different decision makers have different knowledge structures, judgment levels, and personal preferences, which leads to different preference styles used in the decision-making process, such as using a scale of 1–9 or a scale of 0.1–0.9. Then, a hybrid comparison matrix that contains two kinds of judgment in scale form may be obtained. In addition, it is hard to use a set of precise preference styles to describe the complex issues in some cases because of the complexity of decision-making problems and the uncertainty and ambiguity of human thinking. In a word, using the uncertainty judgment style is more appropriate and practicable than the certainty style. As a result, a new expression form of uncertainty called the *hybrid uncertainty comparison matrix* can be obtained for a particular decision-making case. Following that, the methods to express the consistency and carry on decision making based on it are suggested. Then, the application steps suggested in this chapter are explained using an example.

6.1.2 Main Results

6.1.2.1 Related Definitions

Definition 6.1

The value of $a_{ij}{}^L$ is the lower possible value of the judgment, and the value of $a_{ij}{}^U$ is the upper possible value. If the relationship is satisfied, the matrix $\bar{A} = ([a_{ij}{}^L, a_{ij}{}^U])_{n \times n}$ is called a *hybrid uncertain comparison matrix*, where: $a_{ij}{}^L, a_{ij}{}^U \in [1/9, 9]$, $i, j \in S$ and $a_{ij}{}^L, a_{ij}{}^U \in [0.1, 0.9]$, $i, j \notin S$, $S = \{1, \ldots, n\}$. Entries the matrix hold reciprocal and complementary characteristics. That is,

$$i, j \in S, \overline{a_{ji}} = \left[\frac{1}{a_{ij}{}^U}, \frac{1}{a_{ij}{}^L} \right]$$

and

$$i, j \notin S, \overline{a_{ji}} = \left[1 - a_{ij}^{U}, 1 - a_{ij}^{L}\right].$$

The consistency definition of the hybrid uncertain comparison matrix is developed based on the definition of the interval numbers reciprocal comparison matrix and interval numbers complementary comparison matrix.

Definition 6.2

The set w_i, $i = 1, \ldots, n$ is said to be the "weight" of a hybrid uncertain comparison matrix. If the following formula is satisfied, a hybrid uncertain comparison matrix is completely consistent, where:

$$\forall i, j \in S, a_{ij}^{L} \leq \frac{w_i}{w_j} \leq a_{ij}^{U}$$

and

$$\forall i, j \notin S, a_{ij}^{L} \leq \frac{1 + w_i - w_j}{2} \leq a_{ij}^{U}.$$

If the weight w_i, $i = 1, \ldots, n$ does not satisfy the formula, it is not completely consistent.

The complementary consistency of the comparison matrix is given based on the additive consistency definition. If the consistency definition about the complementary judgment element is multiplicative consistency where:

$$\forall i, j \notin S, a_{ij}^{L} \leq \frac{w_i}{w_i + w_j} \leq a_{ij}^{U},$$

the analysis method is similar. If there is certainty judgment in the comparison matrix using a 1–9 scale or a 0.1–0.9 scale, it can be turned into interval number form as $\overline{a_{ij}} = [a_{ij}, a_{ij}]$. Then the weight solution method of hybrid uncertain comparison matrix is researched as follows.

6.1.2.2 Weight Solution Model If a hybrid uncertain comparison matrix is completely consistent, for $a_{ij} \in [1/9, 9]$, $i, j \in S$, Formula (6.1) is satisfied according to Definition 6.2.

$$a_{ij}{}^L \leq \frac{w_i}{w_j} \leq a_{ij}{}^U \ i, j \in S \tag{6.1}$$

Because of the complexity of decision making, and the fuzzy thinking of human judgment, the hybrid comparison matrix is not completely consistent in some situations. Some deviation is permitted and the fuzzy inequality Formula (6.2) is satisfied. The symbol $\tilde{\leq}$ means "fuzzy less than or equal to."

$$a_{ij}{}^L \tilde{\leq} \frac{w_i}{w_j} \tilde{\leq} a_{ij}{}^U, i, j \in S \tag{6.2}$$

Formula (6.2) can be translated into Formula (6.3).

$$\begin{cases} a_{ij}{}^L w_j - w_i \tilde{\leq} 0 \\ w_i - a_{ij}{}^U w_j \tilde{\leq} 0 \end{cases}, i, j \in S \tag{6.3}$$

For $i, j, \notin S$, $a_{ij}{}^L, a_{ij}{}^U \in [0.1, 0.9]$, Formula (6.4) is satisfied when the hybrid uncertain comparison matrix is completely consistent,

$$a_{ij}{}^L \leq \frac{1 + w_i - w_j}{2} \leq a_{ij}{}^U, i, j \notin S \tag{6.4}$$

If the hybrid uncertain comparison matrix is not completely consistent, Formula (6.4) is satisfied in a certain decision-making deviation condition, and then Formula (6.5) is satisfied.

$$a_{ij}{}^L \tilde{\leq} \frac{1 + w_i - w_j}{2} \tilde{\leq} a_{ij}{}^U, i, j \notin S \tag{6.5}$$

Formula (6.5) can be transformed into Formula (6.6).

$$\begin{cases} 2a_{ij}{}^L - w_i + w_j - 1 \tilde{\leq} 0 \\ w_i - w_j + 1 - 2a_{ij}{}^U \tilde{\leq} 0 \end{cases}, i, j \notin S \tag{6.6}$$

Formulas (6.3) and (6.6) can be merged into Formula (6.7).

$$\begin{cases} \begin{cases} a_{ij}{}^{L} w_j - w_i \tilde{\leq} 0 \\ w_i - a_{ij}{}^{U} w_j \tilde{\leq} 0 \end{cases}, i, j \notin S \\ \begin{cases} 2a_{ij}{}^{L} - w_i + w_j - 1 \tilde{\leq} 0 \\ w_i - w_j + 1 - 2a_{ij}{}^{U} \tilde{\leq} 0 \end{cases}, i, j \notin S \end{cases} \tag{6.7}$$

Formula (6.7) can be transformed into Formulas (6.8) and (6.9).

$$\begin{cases} R_s w \tilde{\leq} 0, & (6.8) \\ R_{\bar{s}} w \tilde{\leq} 0, & (6.9) \end{cases}$$

Each line of Formulas (6.8) and (6.9) corresponds to a fuzzy set. The functions μ_s and $\mu_{\bar{s}}$ of the fuzzy set mean the satisfaction degree of line i. Generally, it should satisfy both functions. If the formula is completely unsatisfied, $\mu_s = 0$ and $\mu_{\bar{s}} = 1$ can be obtained. If the formula is fully satisfied, $\mu_s = 1$ and $\mu_{\bar{s}} = 1$ can be obtained. The values of μ_s and $\mu_{\bar{s}}$ are from 0 to 1 and depend on the degree of satisfaction of the formula. As a result, the following linear expression form of μ_s and $\mu_{\bar{s}}$ can be defined as

$$\mu_s = \begin{cases} 1, & R_s w \leq 0 \\ 1 - \dfrac{R_s w}{\delta_s}, & 0 < R_s w \leq \delta_s \\ 0, & R_s w > \delta_s \end{cases}$$

and

$$\mu_{\bar{s}} = \begin{cases} 1, & R_{\bar{s}} w \leq 0 \\ 1 - \dfrac{R_{\bar{s}} w}{\delta_{\bar{s}}}, & 0 < R_{\bar{s}} w \leq \delta_{\bar{s}} \\ 0, & R_{\bar{s}} w > \delta_{\bar{s}} \end{cases}.$$

In the above definition, δ_s and $\delta_{\bar{s}}$ represent the acceptable lack of satisfaction set by decision makers. Because the reciprocal comparison

matrix is based on a scale of 1 to 9 and the complementary comparison matrix is based on a scale of 0.1 to 0.9, the value of δ_S is not equal
to $\delta_{\bar{S}}$. Then, the following method is proposed to set up the acceptable
unsatisfied degree as Formula (6.10).

$$\delta_S = \left(a_{ij}^{\ U} - a_{ij}^{\ L}\right)\xi,\ i,\ j \in S;\ \delta_{\bar{S}} = \left(a_{ij}^{\ U} - a_{ij}^{\ L}\right)\xi,\ i,\ j \notin S \qquad (6.10)$$

In Formula (6.10), $\xi > 0$ and it is a constant. Moreover, the accuracy of decision making will be reduced with the increasing of ξ. In
general, the value of ξ can be increased gradually and the upper limit
can be given according to the degree of awareness of the problem by
decision makers. If the decision makers have the clear judgment level
to the decision-making problem, the value of δ can also be given
directly.

Then, the degree of satisfaction that satisfies Formula (6.7) can be
defined as Formula (6.11).

$$\mu = \min\ (\mu_S,\ \mu_{\bar{S}}) \qquad (6.11)$$

If one needs a definite strategy, one method is to choose the weight
$w_i,\ i = 1, \ldots, n$, which lets Formula (6.11) be the maximum. Introduce
the new variable λ where $0 \leq \lambda \leq 1$ and $\lambda = \mu$, and λ is the degree
that meets Formula (6.7). $\lambda \leq \mu_S$ and $\lambda \leq \mu_{\bar{S}}$ are satisfied. Moreover,
the bigger the value of λ, the better the decision-making result.
Following that, the linear programming model $P_{6.1}$ can be obtained
as Formula (6.12).

$$\max \lambda$$

$$P_{6.1}: \text{s.t.} \begin{cases} \lambda\delta_S + R_S w \leq \delta_S \\ \lambda\delta_{\bar{S}} + R_{\bar{S}} w \leq \delta_{\bar{S}} \\ 0 \leq \lambda \leq 1, w \geq 0 \end{cases} \qquad (6.12)$$

The formula $R_S w$ and $R_{\bar{S}} w$ can be substituted, and the normalized
condition of the weight $\sum_{i=1}^{n} w_i = 1$ should also be considered. Then
the linear programming model $P_{6.2}$ can be obtained.

max λ

$$P_{6.2} : \text{s.t.} \begin{cases} \left(a_{ij}^U - a_{ij}^L\right)\xi\lambda + a_{ij}^L w_j - w_i \leq \left(a_{ij}^U - a_{ij}^L\right)\xi, \, i, \, j \in S \\ \left(a_{ij}^U - a_{ij}^L\right)\xi\lambda + w_i - a_{ij}^U w_j \leq \left(a_{ij}^U - a_{ij}^L\right)\xi, \, i, \, j \in S \\ \left(a_{ij}^U - a_{ij}^L\right)\xi\lambda + 2a_{ij}^L - w_i + w_j - 1 \leq \left(a_{ij}^U - a_{ij}^L\right)\xi, \, i, \, j \notin S \\ \left(a_{ij}^U - a_{ij}^L\right)\xi\lambda + w_i - w_j + 1 - 2a_{ij}^L \leq \left(a_{ij}^U - a_{ij}^L\right)\xi, \, i, \, j \notin S \\ \sum_{i=1}^n w_i = 1, \, 0 \leq \lambda \leq 1, \, w_i \geq 0, \, i = 1, \dots, n \end{cases}$$

(6.13)

Let the optimal solution of $P_{6.2}$ be λ^*. Theorem 6.1 can be obtained.

Theorem 6.1

When $\lambda^* = 1$, the hybrid uncertain comparison matrix is completely consistent and vice versa.

Proof

When $\lambda^* = 1$, $w_i = 1, \dots, n$ can be obtained from model $P_{6.2}$. At the same time, it meets the following formula:

$$\begin{cases} \left(a_{ij}^U - a_{ij}^L\right)\xi\lambda + a_{ij}^L w_j - w_i \leq \left(a_{ij}^U - a_{ij}^L\right)\xi, \, i, \, j \in S \\ \left(a_{ij}^U - a_{ij}^L\right)\xi\lambda + w_i - a_{ij}^U w_j \leq \left(a_{ij}^U - a_{ij}^L\right)\xi, \, i, \, j \in S \\ \left(a_{ij}^U - a_{ij}^L\right)\xi\lambda + 2a_{ij}^L - w_i + w_j - 1 \leq \left(a_{ij}^U - a_{ij}^L\right)\xi, \, i, \, j \notin S \\ \left(a_{ij}^U - a_{ij}^L\right)\xi\lambda + w_i - w_j + 1 - 2a_{ij}^L \leq \left(a_{ij}^U - a_{ij}^L\right)\xi, \, i, \, j \notin S \end{cases}$$

and Formulas (6.1) and (6.4) are satisfied. According to Definition 6.2, the hybrid uncertain comparison matrix is completely consistent. Otherwise, when the hybrid uncertain comparison matrix is completely consistent, both Formulas (6.1) and (6.4) are also satisfied. This means that $w_i, \, i = 1, \dots, n$ also meets the condition of

$$\begin{cases} a_{ij}{}^{L}w_j - w_i \leq 0, \ i, \ j \in S \\ w_i - a_{ij}{}^{U}w_j \leq 0, \ i, \ j \in S \\ 2a_{ij}{}^{L} - w_i + w_j - 1 \leq 0, \ i, \ j \notin S \\ w_i - w_j + 1 - 2a_{ij}{}^{L} \leq 0, \ i, \ j \notin S \end{cases}.$$

As a result, the value of λ^* equals 1.

Theorem 6.2

When $\lambda^* < 1$, the hybrid uncertain comparison matrix is not completely consistent. The smaller λ^* is, the worse the consistency is.

Proof

w_i Does not satisfy the condition of

$$\begin{cases} a_{ij}{}^{L}w_j - w_i \leq 0, \ i, \ j \in S \\ w_i - a_{ij}{}^{U}w_j \leq 0, \ i, \ j \in S \\ 2a_{ij}{}^{L} - w_i + w_j - 1 \leq 0, \ i, \ j \notin S \\ w_i - w_j + 1 - 2a_{ij}{}^{L} \leq 0, \ i, \ j \notin S \end{cases}$$

as $\lambda^* < 1$. So the hybrid uncertain comparison matrix is not completely consistent. In addition, the smaller the value of λ^*, the lower the degree that meets Formula (6.7). That is, the consistency is worse.

According to Theorem 6.1 and Theorem 6.2, λ^* can be considered the consistency degree index of the hybrid uncertain comparison matrix.

Definition 6.3

λ^* can be called the consistency degree of a hybrid uncertain comparison matrix as there is an optimal solution or its feasible region is empty of $P_{6.2}$.

If there is an empty feasible region in $P_{6.2}$, $\lambda^* = 0$ can be obtained. The comparison matrix is of unacceptable consistency for decision makers. The smaller the value of λ^*, the lower the comparison matrix's consistency. For example, $\lambda^* < 0.6$ means that the degree of satisfaction given by experts is smaller than 0.6 or the feasible region of $P_{6.2}$ is empty. This means that the decision makers' logic is quite chaotic or there are serious differences among the experts' understandings. In this case, the experts need to check their judgments, communicate with each other, and make some improvements. Then the suggested model can be applied.

In general, there may be many sets of optimal solutions of model $P_{6.2}$. Therefore, the solution is not exclusive, and therein may lie a misleading problem. The two-stage method for the weight is used to solve this problem. The value of λ^* is used to construct the new weight solution model. Then, the distribution value of w_i based on the consistency degree λ^* can be obtained. Model $P_{6.3}$ shows the process of solving the two-stage method.

$$\max/\min w_i, \ i = 1, \ldots, n$$

$$P_{6.3}: \text{s.t.} \begin{cases} \lambda = \lambda^* \\ \left(a_{ij}^U - a_{ij}^L\right)\xi\lambda + a_{ij}^L w_j - w_i \leq \left(a_{ij}^U - a_{ij}^L\right)\xi, \ i, j \in S \\ \left(a_{ij}^U - a_{ij}^L\right)\xi\lambda + w_i - a_{ij}^U w_j \leq \left(a_{ij}^U - a_{ij}^L\right)\xi, \ i, j \in S \\ \left(a_{ij}^U - a_{ij}^L\right)\xi\lambda + 2a_{ij}^L - w_i + w_j - 1 \leq \left(a_{ij}^U - a_{ij}^L\right)\xi, \ i, j \notin S \\ \left(a_{ij}^U - a_{ij}^L\right)\xi\lambda + w_i - w_j + 1 - 2a_{ij}^L \leq \left(a_{ij}^U - a_{ij}^L\right)\xi, \ i, j \notin S \\ \sum_{i=1}^n w_i = 1, \ w_i \geq 0, \ i = 1, \ldots, n \end{cases}$$

$$(6.14)$$

If the value of ξ is given and there is no feasible region of $P_{6.2}$ or $P_{6.3}$, the value ξ has to be adjusted to a higher value. At the same time, the accuracy of the decision making will be decreased. When ξ is large, the scope of deviation can be obtained from Formula (6.10). In this case, one can confirm that decision makers' judgments are not consistent. The comparison needs to be reexamined and the corresponding judgment will be made.

The weight w_i can be shown the interval numbers as $w_i = [w_i^L, w_i^U]$, $i = 1, \ldots, n$ in condition of λ^* after solving model $P_{6.3}$. The decision makers can rank with the interval weight to obtain the ultimate order.

6.1.2.3 Sorting Method of Interval Numbers Weight

Definition 6.4

Set two interval numbers $a = [a^L, a^U]$ and $b = [b^L, b^U]$; let $l_a = a^U - a^L$ and $l_a = b^U - b^L$. Formula (6.15) is called the possibility degree of $a \geq b$ [60].

$$p(a \geq b) = \frac{\max\{0, l_a + l_b - \max(b^U - a^L, 0)\}}{l_a + l_b} \tag{6.15}$$

Based on Definition 6.4, we can obtain the probability degree and establish the probability matrix to the weight $w_i = [w_i^L, w_i^U]$, $i = 1, \ldots, n$ obtained based on the hybrid uncertain comparison matrix. This matrix contains the information $w_i > w_j$, which means that the possibility of $w_i > w_j$ is p_{ij}. In addition, the ranking interval number weight can be transformed into comparing the possibility degree matrix. One can obtain the weight vector w''_i, $i = 1, \ldots, n$, which means the priority of the weight. Finally, according to w''_i, the ultimate ranking based on possibility can be obtained.

6.1.3 Example Analysis

A venture capital company has funds to carry on the optimal investment. There are four options: a biopharmaceutical company, a food company, a fashion company, and a computer software company. Suppose that the hybrid uncertain comparison matrix given by experts is as follows:

$$\begin{bmatrix} [1, 1] & [2, 4] & [0.5, 0.6] & [1/2, 1] \\ [1/4, 1/2] & [1, 1] & [0.4, 0.5] & [1/6, 1/5] \\ [0.4, 0.5] & [0.5, 0.6] & [1, 1] & [0.3, 0.5] \\ [1, 2] & [5, 6] & [0.5, 0.7] & [1, 1] \end{bmatrix}$$

Suppose that $\xi = 10\%$, then $\lambda^* = 1$ can be obtained by solving model $P_{6.2}$. This means that the judgment given by decision makers is completely consistent.

In addition, the weight of $\overline{w_1} = [0.227, 0.343]$, $\overline{w_2} = [0.067, 0.104]$, $\overline{w_3} = [0.1, 0.275]$, $\overline{w_4} = [0.364, 0.525]$ can be obtained according to model $P_{6.3}$. The possibility degree matrix

$$\begin{bmatrix} 0.5 & 1 & 0.835 & 0 \\ 0 & 0.5 & 0.02 & 0 \\ 0.165 & 0.98 & 0.5 & 0 \\ 1 & 1 & 1 & 0.5 \end{bmatrix}$$

can be obtained according to Formula (6.15). The weight is as follows: that is, $w_1 = 0.292$, $w_2 = 0.065$, $w_3 = 0.206$, $w_4 = 0.438$. The ultimate order is $w_4 > w_1 \overset{1}{>} w_3 \overset{0.835}{>} w_2$. From the result, this venture capital should be invested in a computer software company.

6.2 Research on Four Kinds of Uncertain Preference Information Aggregation Approaches in Group Decision Making

6.2.1 Background

With the development of society and progress in science and technology, knowledge and information have increased greatly. Decision-making problems are becoming more complex. Further, the application of the group decision-making process is widespread. In the group decision-making process, the different preference information structures of comparison matrix, utility value, and preference order are adopted due to the decision-making difference in the knowledge structures, individual preferences, and judgment levels. In this case, the aggregation approach of how to aggregate the single decision maker into the group preference is studied. According to the preference structure in the group decision-making process, the aggregation approaches are divided into the same structure aggregation approach and the different structure aggregation approach. The same preference structure aggregation method has already obtained many research results [47, 57–53], whereas the different preference structure aggregation research is still a new topic. As a result of the complexity and uncertainty of decision problems and the fuzziness of human thinking, it is not realistic to use the certain preference information in the complex decision-making process. In fact, the uncertainty is absolute and the certainty is relative. The literature has summarized the progress in the field of uncertainty decision making. The available research is limited only to the single preference information. The consistency and weight estimations of the single uncertainty preference structure are studied, while the different structure aggregation has received little attention. Based on four kinds of group decision-making methods of the uncertain preference information, an aggregation approach is put forward in this chapter and the consistency level of the group is defined.

6.2.2 Major Results

6.2.2.1 Basic Definition

Definition 6.5

Set the decision maker aiming at projects set X to give a comparative judgment matrix, $\bar{A} = (\bar{a}_{ij})_{n \times n}$ for interval number reciprocal judgment matrices, $\bar{a}_{ij} = [a_{ij}^L, a_{ij}^U]$, $a_{ij}^L \leq a_{ij}^U$, a_{ij}^U, for a_{ij}s upper limit, a_{ij}^L for a_{ij}s lower limit. $\bar{a}_{ji} = [1/a_{ij}^U, 1/a_{ij}^L]$, $\bar{a}_{ii} = [1, 1]$.

Definition 6.6

Set the decision maker aiming at project set X to give comparisons of the project's preference relation. The judgment matrix $\bar{B} = (\bar{b}_{ij})_{n \times n}$ for interval number complementary judgment matrices, $\bar{b}_{ij} = [b_{ij}^L, b_{ij}^U]$, $b_{ij}^L \leq b_{ij}^U$, $\bar{b}_{ji} = [1 - b_{ij}^U, 1 - b_{ij}^L]$, $\bar{b}_{ii} = [0.5, 0.5]$.

Definition 6.7

Set the decision maker aiming at project set X to give preference ordering $\bar{F}_i = [f_i^L, f_i^U]$ $f_i^L, f_i^U \geq 0$ as the integral type number. The value of f_i means projects x_is position and sequence in all program, $f_i^L \leq f_i^U$. Generically, the lower preference ordering f_i is, the better corresponding program is.

Definition 6.8

Set the decision maker aiming at project set X to give the value of utility $\bar{E}_i = [e_i^L, e_i^U]$ $e_i^L, e_i^U \geq 0$ is a real number, $e_i^L \leq e_i^U$. The greater the value of utility e_i is, the better the corresponding program is. The value of e_i can be regarded as program x_i weight.

6.2.2.2 Aggregation Model

The literature has reported the weight approach of interval number reciprocal judgment matrices and interval number complementary judgment matrices. But there is still no literature to establish a solving weight model to solve the uncertainty of preference information. If each kind of uncertain preference information uses the respective weight solution method, there is no doubt that it will increase the difficulty of group decision making. Also, because each kind of weight solution method mechanism was certainly not the same, it is hard to ensure the reliability of aggregation. In addition, the very uncertain preference information weight is still presented in uncertain form; this also increase the difficulty of the result analysis of group decision making. Based on the existing research [57, 58, 60, 71], this chapter analyzes the characteristics of uncertain preference information and establishes a unified model of solving each uncertain preference information weight, then proposes a method of building many kinds of uncertain preference information.

1. Weights modeling for interval number reciprocal judgment matrices: For $\overline{A} = (\overline{a_{ij}})_{n \times n}$, the result of its weight is wa_i, $i = 1, \ldots, n$; generally, if

$$a_{ij}^{L} \leq \frac{wa_i}{wa_j} \leq a_{ij}^{U}, \forall i, j \tag{6.16}$$

$\overline{A}$ has complete consistency. If $\overline{A}$ does not have complete consistency, introduce deviation values of random variables ap_{ij}, ad_{ij}, so Formula (6.17) is tenable.

$$\begin{cases} a_{ij}^{L} wa_j \leq wa_i + ap_{ij}, i, j = 1, \ldots, n, i \neq j \\ wa_i \leq a_{ij}^{U} wa_j + ad_{ij}, i, j = 1, \ldots, n, i \neq j \end{cases} \tag{6.17}$$

The lower the deviation values ap_{ij}, ad_{ij} are, the better $\overline{A}$'s complete consistency is. If $ap_{ij} = 0$, $ad_{ij} = 0$, $\forall i, j$, and $\overline{A}$ has complete consistency, we can set model $P_{6.4}$ to evaluate the weight for interval number reciprocal judgment matrices.

$$\min \sum_{i,j} ap_{ij} + ad_{ij} \tag{6.18}$$

$$\text{s.t.} \begin{cases} a_{ij}^{L} wa_j \leq wa_i + ap_{ij}, i, j = 1, \ldots, n, i \neq j & (6.19) \\ wa_i \leq a_{ij}^{U} wa_j + ad_{ij}, i, j = 1, \ldots, n, i \neq j & (6.20) \\ \sum_{i=1}^{n} wa_i = 1 & (6.21) \\ wa_i \geq 0, ap_{ij}, ad_{ij} \geq 0, i = 1, \ldots, n & (6.22) \end{cases}$$

Formula (6.18) is expressed to look for a group of weights wa_i and minimize the sum of the deviation values of random variables. Formulas (6.19)–(6.20) are the same as Formula (6.17); Formula (6.21) expresses the weight that satisfies normalization condition. Formula (6.22) is for weight wa_i and the deviation values that are both in need of nonnegative restriction.

2. Weights modeling for interval number complementary judgment matrices: For interval number complementary judgment matrix $B = (b_{ij})_{n \times n}$ the result of its weight is wb'_i. If it has complete consistency, for

$$\forall i, j \; b_{ij} = \frac{wb'_i}{wb'_i + wb'_j}.$$

According to the consistency formula of interval number complementary matrix, Formula (6.16), for $\bar{B} = (\bar{b}_{ij})_{n \times n}$ the result of its weight is wb'_i. If it has complete consistency, then

$$b_{ij}{}^L \leq \frac{wb_i}{wb_i + wb_j} \leq b_{ij}{}^U \tag{6.23}$$

If $\bar{B}$ does not have complete consistency, introduce deviation variable bp_{ij}, bd_{ij}, so one can obtain

$$\begin{cases} b_{ij}{}^L (wb_i + wb_j) \leq wb_i + bp_{ij} \\ wb_i \leq b_{ij}{}^U (wb_i + wb_j) + bd_{ij} \end{cases} \tag{6.24}$$

The lower bp_{ij}, bd_{ij} is, the better the consistency of $\bar{B}$ is. If $bp_{ij} = 0$, $bd_{ij} = 0$, $\forall i, j$, $\bar{B}$ has complete consistency, and we can set $P_{6.5}$ to estimate the weight for interval number complementary judgment matrices.

$$\min \sum_{i,j} bp_{ij} + bd_{ij} \tag{6.25}$$

$$\text{s.t.} \begin{cases} b_{ij}{}^L (wb_i + wb_j) \leq wb_i + bp_{ij}, i \neq j & \tag{6.26} \\ wb_i \leq b_{ij}{}^U (wb_i + wb_j) + bd_{ij}, i \neq j & \tag{6.27} \\ \sum_{i=1}^{n} wb_i = 1 & \tag{6.28} \\ wb_i \geq 0, bp_{ij}, bd_{ij} \geq 0, i = 1, \ldots, n & \tag{6.29} \end{cases}$$

Formula (6.25) is expressed to look for a group of weights wb_i and minimize the sum of the deviation values. Formulas (6.26)–(6.27) are the same as Formula (6.24); Formula (6.28) expresses the weight satisfied normalization condition. Formula (6.29) is for weight wa_i and the deviation values that are both in need of nonnegative restriction.

3. Weights modeling for the value of utility of interval number: If we use the value of utility of interval number $\bar{E}_i = [e_i^L, e_i^U]$ to express its preference, we can consider it as the preferential weight of the program. Set the real weight for we_i, and $e_i^L \le we_i \le e_i^U$. Introduce deviation values ep_i, ed_i; then we have Formula (6.30).

$$e_i^L \le we_i + ep_i, \quad we_i \le e_i^U + ed_i \tag{6.30}$$

The lower the ep_i, ed_i in Formula (6.30) is, the better the consistency of decision making is. So we can set model $P_{6.6}$ to solve the weight.

$$\min \sum_i ep_i + ed_i \tag{6.31}$$

$$\text{s.t.} \begin{cases} e_i^L \le we_i + ep_i, i = 1, \ldots, n & (6.32) \\ we_i \le e_i^U + ed_i, i = 1, \ldots, n & (6.33) \\ \sum_i we_i = 1, we_i \ge 0 & (6.34) \end{cases}$$

Formula (6.31) is expressed to look for a group of weights we_i and minimize the sum of deviation values of random variables. Formulas (6.32)–(6.33) are the same as Formula (6.30); Formula (6.34) expresses the weight satisfied normalization condition and nonnegative restriction.

4. Weights modeling for interval number preference sequence: If we use interval number preference sequence $\bar{F}_i = [f_i^{\prime L}, f_i^{\prime U}]$ to express preference, setting the weight of the last program is a positive parameter. This chapter proposes to change the interval number preference sequence into the value of utility $[f_i^L, f_i^U]$; that is,

$$f_i^L = \frac{\dfrac{n - f_i'^U}{n-1} + \varepsilon}{\sum \dfrac{n - f_i'^U}{n-1} + \varepsilon}$$

$$\Rightarrow \frac{\dfrac{n - f_i'^U}{n-1} + \varepsilon}{\dfrac{n}{2} + n\varepsilon}, \; f_i^U = \frac{\dfrac{n - f_i'^L}{n-1} + \varepsilon}{\dfrac{n}{2} + n\varepsilon} \tag{6.35}$$

$\varepsilon > 0$, which is a positive parameter. If the weight of the last program is x, we can obtain

$$\varepsilon = \frac{nx}{2(1 - nx)}$$

from Formula (6.35). Therefore, we can use the method of the value of utility of interval number to solve the problem of interval order, which is $P_{6.7}$. The parameter implication of $P_{6.7}$ is the same as $P_{6.6.}$

$$P_{6.7} : \min \sum_i f p_i + f d_i$$

$$\text{s.t.} \begin{cases} \dfrac{\dfrac{n - f_i^U}{n-1} + \varepsilon}{\dfrac{n}{2} + n\varepsilon} \le w f_i + f p_i, \, i = 1, \ldots, n \\[2em] w f_i \le \dfrac{\dfrac{n - f_i^L}{n-1} + \varepsilon}{\dfrac{n}{2} + n\varepsilon}, \, f d_i, \, i = 1, \ldots, n \\[2em] \sum_i w f_i = 1, \, w f_i \ge 0 \end{cases}$$

5. Aggregation model for the general idea of group decision making: In group decision making, many decision makers have proposed the above-mentioned uncertain preference

styles or their various combinations. The following text will study how to aggregate so much preference information to obtain the group preferences.

Generally, decision makers in a group should reach a coincident opinion, which is expressed by w_i, $i = 1, \ldots, n$. Set the given expert weight of interval number reciprocal judgment matrices p_1, the expert weight of interval number complementary judgment matrices p_2, the expert weight of the value of utility of interval number p_3, and the expert weight of interval number preference sequence p_4, so there is an aggregation model $P_{6.8}$.

$$P_{6.8} : \min \sum_{i,j} p_1(ap_{ij} + ad_{ij}) + p_2(bp_{ij} + bd_{ij}) +$$

$$\sum_i p_3(ep_i + ed_i) + p_4(fp_i + fd_i) \tag{6.36}$$

$$\text{s.t.} \begin{cases} a_{ij}^L w_j \leq w_i + ap_{ij}, \\[2mm] w_i \leq a_{ij}^U w_j + ad_{ij}, i \neq j = 1, \ldots, n & (6.37) \\[2mm] b_{ij}^L(w_i + w_j) \leq w_i + bp_{ij}, \\[2mm] w_i \leq b_{ij}^U(w_i + w_j) + bd_{ij}, i \neq j = 1, \ldots, n & (6.38) \\[2mm] e_i^L \leq w_i + ep_i, w_i \leq e_i^U + ed_i, i = 1, \ldots, n & (6.39) \\[2mm] \dfrac{\dfrac{n - f_i^U}{n-1} + \varepsilon}{\dfrac{n}{2} + n\varepsilon} \leq w_i + fp_i, w_i \leq \dfrac{\dfrac{n - f_i^L}{n-1} + \varepsilon}{\dfrac{n}{2} + n\varepsilon} + fd_i, i = 1, \ldots, n & (6.40) \\[4mm] \displaystyle\sum_{i=1}^n w_i = 1, w_i \geq 0 & (6.41) \\[2mm] ap_{ij}, ad_{ij}, bp_{ij}, bd_{ij}, ep_i, ed_i, fp_i, fd_i \geq 0, i, j = 1, \ldots, n & (6.42) \end{cases}$$

In model $P_{6.8}$, the sum of restraint deviation variable that is minimally expressed by the objective function (6.36), Formula

(6.37) has the same meaning as Formula (6.17), Formula (6.38), and Formula (6.24). Formula (6.39) and Formula (6.30) also have the same meaning. Formula (6.40) and model $P_{6.7}$ have the same constraint condition. Formula (6.41) means that weights satisfied the condition of normalization and nonnegativity. Formula (6.42) means that the deviation variable satisfied the condition of nonnegativity.

If the information given by the expert includes both certainty preference information and uncertainty preference information, then certainty preference information will be changed into uncertainty preference information; for example, carrying on the element $a_{ij} \rightarrow [a_{ij}, a_{ij}]$, then model $P_{6.8}$ can be used to solve the problem. Model $P_{6.8}$ is the linear programming model and the scale of the solution is not large, so we may use the general linear programming software to carry on.

Theorem 6.3
Model $P_{6.8}$ must have the optimal solution.

Proof
It is easy to know how, by adjusting the sizes of various deviations, the weights satisfying (6.37)–(6.41) can be obtained. Therefore, model $P_{6.8}$ must have the optimal solution.

Theorem 6.3
Recording θ^* is the optimal solution in model $P_{6.8}$; if $\theta^* = 0$, the decision-making group's opinion is completely consistent. If the opinion is incompletely consistent, then it must be $\theta^* > 0$. Also, if θ^* is greater the group opinion is more dispersible.

Proof
θ^* is composed by the nonnegative deviation variable. If $\theta^* = 0$ and $\exists i, j, ap_{ij}, ad_{ij}, bp_{ij}, bd_{ij}, ep_i, ed_i, fp_i, fd_i = 0$, then (6.37)–(6.41) are

established. Therefore, each expert's opinion is completely consistent. If $\theta^* > 0$, then ap_i, ad_i, bp_i, bd_i, ep_i, ed_i, fp_i, fd_i must have one for non-zero. Therefore, the decision-making group's opinion is incompletely consistent. In addition its value is greater, formulas (6.37)–(6.41) conform to the degree to be worse; the community opinion also jumps over for dispersible.

According to Theorem 6.3, through solving model $P_{6.8}$, synthesis preference information, which gives many kinds of uncertainty information situations in the policy-making community, can be obtained. The following questions still need to be further considered:

1. Synthesis preference always can obtain in a decision-making community according to model $P_{6.8}$. But are numerous position experts' opinions consistent? If consistent, how to express the degree of uniformity? If inconsistent, how is the degree of divergence?

2. How do we determine the experts' weight in model $P_{6.8}$? Usually, the value of experts' weight is very difficult to determine. If various experts' weights are equal, then the obtained result is the expert advice compromise, which does not have the consideration commonly used "most" in the decision-making principle—whether experts' weight carries on the suitable evaluation according to the decision-making community's synthesis by chance. In question 1, the size of θ^* can be used to determine the degree of uniformity of the decision-making group. However, what size of θ^* can express the decision-making group's degree of uniformity as good and the expert opinion as consistent? Below, the inconsistent degree size of the expert advice is analyzed through the vector method.

Theorem 6.4

When $p_1 = 1$, p_2, p_3, $p_4 = 0$, the decision-making group's synthesis preference obtained through model $P_{6.8}$, w_i, $i = 1, \ldots, n$ completely obeys expert 1's opinion. When $p_2 = 1$, p_1, p_3, $p_4 = 0$, the results of $P_{6.8}$ obey expert 2's opinions completely. The rest can be done in the same manner.

Proof

When $p_1 = 1$, p_2, p_3, $p_4 = 0$, according to model $P_{6.8}$, the value of synthesis preference was restricted by the constraint condition formulas (6.37)–(6.41) deviation variable; it is always established by adjusting other constraint conditions in the formula the deviation variable. Therefore, the decision-making group's synthesis preference is decided by formulas (6.37)–(6.41). Other situations' proofs are similar.

Based on model $P_{6.8}$, set p_1, p_2, p_3, p_4 equal to 1 separately, and the three others are equal to 0. Theorem 6.4 is obeyed completely in this expert opinion, recording w_i^k, $i = 1, \dots, n$. According to the vector inner product formula, the angle of the vector has reflected the consistency degree between two vectors; therefore, based on w_i, w_i^k calculated separately, expert k and the synthesis opinion uniformity conforms to the consistency degree. If expert k and the synthesis opinion uniformity conformity to the degree is good, then this expert's weight should also be supposed to be larger; otherwise supposed to be small.

Definition 6.9

Set

$$\eta^k = \frac{w \bullet w^k}{|w||w^k|}$$

to be called expert's judgment uniform level, in which $w = [w_1, \dots, w_n]$, $w^k = [w_1^k, \dots, w_n^k]$ is obtained separately from model $P_{6.8}$.

According to Definition 6.9, $1 \geq \eta^k \geq 0$. $\eta^k = 1$ indicates that expert k's opinion is completely consistent with the decision-making group's synthesis preference while $\eta_k = 0$ indicates that the expert opinion violates the synthesis preference of the decision-making community completely. The smaller the value of η^k is, the greater the expert's opinion and the decision-making group's comprehensive opinion deviation is.

Based on Definition 6.9, expert k's weight is defined

$$p_k = \frac{\eta^k}{\sum_k \eta^k}.$$

Based on p_k recomputing model $P_{6.8}$, the new result w_i, $i = 1, \ldots, n$ obtained will take the synthesis preference of the decision-making group.

Definition 6.10

Set

$$\eta = \frac{\sum_{k=1}^{K} \eta^k}{K}$$

to be called the average uniform degree of the decision-making group's opinion. K is the experts' number.

The significance of the vector method is more explicit, compared to the method of θ^*, which determines the consistency degree of the decision-making group's opinion. When the average uniform consistency of the decision-making group's opinion is large the expert opinion is consistent to the question; otherwise the expert opinion is dispersing. In this case, the decision-making group should examine the question carefully, carrying on the appropriate readjustment to their own judgment to achieve a more consistent decision-making result.

The steps of the method in this chapter are stated as follows:

Step 1: Initially supposing that various experts' weights closely coincide. A compromised synthesis preference result can be obtained by the decision-making group though computing model $P_{6.8}$.

Step 2: Calculating expert's judgment uniform level η_k obtains the expert's weight.

Step 3: Recomputing model $P_{6.8}$, the result obtained takes the group finally by chance. The computation obtains the group's average uniform degree.

6.2.3 Example Analysis

A risk investment company wants make an optimal investment. There are four alternatives: a biopharmaceutical company, a food company, a fashion company, and a computer software company. The company employs m experts to make decisions ($m \geq 2$) as well as interval number reciprocal judgment matrices, interval number complementary judgment matrices, interval number preference, and interval number utility values.

$$A = \begin{bmatrix} [1,1] & [1,2] & [2,3] & [1,2] \\ [1/2,1] & [1,1] & [3,5] & [1/2,1] \\ [1/3,1/2] & [1/5,1/3] & [1,1] & [1,2] \\ [1/2,1] & [1,2] & [1/2,1] & [1,1] \end{bmatrix},$$

$$B = \begin{bmatrix} [0.5,0.5] & [0.5,0.6] & [0.6,0.7] & [0.5,0.6] \\ [0.4,0.5] & [0.5,0.5] & [0.5,0.6] & [0.3,0.4] \\ [0.3,0.4] & [0.4,0.5] & [0.5,0.5] & [0.2,0.3] \\ [0.4,0.5] & [0.6,0.7] & [0.7,0.8] & [0.5,0.5] \end{bmatrix}$$

$$f_1 = [0.4, 0.5], f_2 = [0.1, 0.2], f_3 = [0.1, 0.2], f_4 = [0.2, 0.4],$$

$$e_1 = [1, 2], e_2 = [2, 3], e_3 = [2, 4], e_4 = [1, 2].$$

Set $\varepsilon = 0.05$, solve model $P_{6.8}$, and obtain the weight of these four programs. They are $w_1 = 0.375$, $w_2 = 0.25$, $w_3 = 0.125$, $w_4 = 0.25$.

Set the k expert's weight p_k as 1; the others are 0. Solve model $P_{6.8}$ and list the result in Table 6.1. The four experts think program 3 is the worst program, and three experts that think the best program is program 1.

According to Table 6.1, the experts' group consistency level is calculated respectively as: 0.9829, 0.9831, 0.9237, 0.8990. Then the experts' weight is calculated respectively as: 0.259, 0.259, 0.245, 0.237. Recomputing model $P_{6.8}$, the result obtained and the original result are the same. The experts' average consistency degree is 0.947, the expert opinion is consistent. Alternative 1 is the synergy; alternative 2 and alternative 4 are next; and alternative 3 is the most inferior plan.

Table 6.1 Decision-Making Result of Expert and the Group

ALTERNATIVE	EXPERT 1		EXPERT 2		EXPERT 3		EXPERT 4		GROUP	
	WEIGHT	RANK	WEIGHT	RANK	WEIGHT	RANK	WEIGHT	RANK	WEIGHT	RANK
1	0.333	1	0.323	1	0.4	1	0.326	2	0.375	1
2	0.333	1	0.215	2	0.1	2	0.174	3	0.25	2
3	0.111	3	0.138	3	0.1	2	0.023	4	0.125	3
4	0.222	2	0.323	1	0.4	1	0.477	1	0.25	2

6.3 Research on Large-Scale Group Decision
   Approach Based on Gray Cluster

6.3.1 Background

Along with the widespread application of internet technology, it is possible for more decision makers to participate in the decision making in a complex decision-making case. The number of decision makers who participate in some significant decision-making question increases day by day. It is important to study effectively the method of dealing with a large-scale group decision approach. As a result of decision makers' different knowledge structures, judgment levels, and personalities, decision makers may use preference information of multiple different structural forms, such as judgment matrix, utility value, preference ordering, and so on. Because the information for decision making is incomplete and asymmetric, multiple forms of the uncertain preference information are given by decision makers frequently. As a result, in some decision-making cases, using an uncertain form of expression to express the preferences of decision makers is more appropriate and is easily accepted by decision makers. In research on aggregating multiple decision makers' preferences, the aggregated research of multiple structure certainty preference has achieved rich results, while the aggregating research of multiple structure uncertainty preference has not obtained more attention according to the information available to the author. Difficulties lie in the following three aspects to this question.

First, it is hard to aggregate and coordinate a large number of preferences of decision makers. Because of decision makers' different knowledge structure, personal preferences, subjective positions, and so on, decision makers may have inconsistent views. Due to the differences in their views, it is often difficult to coordinate and communicate with a large number of decision makers, and it is impossible to obtain the consistent result of group decision making.

Second, it is hard to identify the key decision-making individual out of a large number of decision makers. During the group decision-making process, give management and control to the key decision-making individual so he or she can coordinate group differences effectively. Decision makers generally have different judgment levels and individual preferences. How to excavate the judgment information of a decision-making individual and discover the key decision-making individual who has the more tremendous influence function to the group preference is quite important.

Third, multiple types of uncertainty preference aggregation methods are lacking. In an open decision-making situation, a large-scale decision-making group may use many kinds of judgment preference forms. In some complex decision-making situations, the determined judgment is often difficult to give by the decision makers. So how to deal with the complex preference forms and make decisions about ambiguous uncertainties judgment is a difficult task?

This chapter mainly concentrates on the following three aspects: First, in view of multiple structure uncertainty judgment preferences that are given by decision makers of a large-scale decision-making group, the ideas of consistent information uncertainty are proposed. Second, use the method of gray clustering to classify the group preferences. It can reduce the coordinated difficulty of the group decision. Based on the group categories of gray clustering, the method of aggregating decision makers' preferences is proposed to obtain a comprehensive preference. Third, propose the search method of the core decision maker to coordinate and communicate to the group. In this foundation, analyze comprehensive preferences of every kind of group and aggregate comprehensive preferences. Then a view of the large-scale decision-making group can be obtained, as shown in Figure 6.1.

6.3.2 Main Results

6.3.2.1 Basis Definition

Definition 6.11

Suppose the decision maker produces a multiple comparison judgment matrix in view of the project collection X, and call $\overline{A} = (\overline{a_{ij}})_{n \times n}$ the reciprocal judgment matrix of interval number. $\overline{a_{ij}} = [a_{ij}^{L}, a_{ij}^{U}]$,

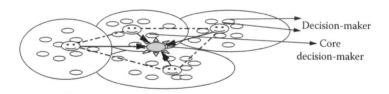

Figure 6.1 The process of clustering and coordinating a large-scale group decision.

$a_{ij}^L \leq a_{ij}^U$, a_{ij}^U is the upper value of a_{ij} and a_{ij}^L is the lower value of a_{ij}. $\overline{a}_{ji} = [1/a_{ij}^U, 1/a_{ij}^L]$, $\overline{a}_{ii} = [1/1]$.

Definitions 6.12

Suppose that the decision maker produces a relationship of fuzzy preference of multiple comparisons in view of the project collection X, and call judgment matrix $\overline{B} = (\overline{b}_{ij})_{n \times n}$ the complementary judgment matrix of interval number, which has

$$\overline{b}_{ij} = [b_{ij}^L, b_{ij}^U], b_{ij}^L \leq b_{ij}^U, \overline{b}_{ji} = [1 - b_{ij}^U, 1 - b_{ij}^L], \overline{b}_{ii} = [0.5, 0.5].$$

Definition 6.13

Suppose that the decision maker produces the value of order relation $\overline{F}_i = [f_i^L, f_i^U]$ in view of the project collection X, which has $f_i^L, f_i^U \geq 0$ and $\overline{F}_i$ is an integer. f_i is the position of x_i in all projects and $f_i^L \leq f_i^U$. In general, the smaller the value of f_i, the better the corresponding project.

Definition 6.14

Suppose that the decision maker produces the utility value $\overline{E}_i = [e_i^L, e_i^U]$ in view of the project collection X, and $e_i^L, e_i^U \geq 0$. It is a real number and $e_i^L \leq e_i^U$. The higher the utility value of e_i, the better the corresponding project. e_i can be supposed as the weight of project x_i.

6.3.2.2 Unified Weight Approach of Multiple Uncertainty Preference Structure There is much research on reciprocal judgment matrices of interval number, complementary judgment matrices of interval number, and other uncertainty preferences weight solution in the literature [65, 66, 83]. But the mechanism of every kind of priority methods is different, so it will increase the difficulty of aggregating a group undoubtedly. On the basis of the existing research [56, 57, 71, 72], this chapter proposes the unified model of uncertainty preference sorting through introducing a deviation value of a random variable.

wa_i, $i = 1, \ldots, n$ can be called the *derived weight* of reciprocal judgment matrix of interval $\overline{A} = (\overline{a}_{ij})_{n \times n}$. If there is

$$a_{ij}^{\ L} \leq \frac{wa_i}{wa_j} \leq a_{ij}^{\ U}, \forall i,j$$

and $a_{ij}^{\ U}$ is the upper value of a_{ij} and $a_{ij}^{\ L}$ is the lower value of a_{ij}, $\bar{A}$ has complete consistency. If $\bar{A}$ is not completely consistent, introduce the deviation value ap_{ij}, ad_{ij}. Formula (6.43) can be obtained.

$$\begin{cases} a_{ij}^{\ L} wa_j \leq wa_i + ap_{ij}, i, j = 1, \ldots, n, i \neq j \\ wa_i \leq a_{ij}^{\ U} wa_j + ad_{ij}, i, j = 1, \ldots, n, i \neq j \end{cases} \tag{6.43}$$

The smaller the value of deviation value ap_{ij}, ad_{ij}, the better the consistency of $\bar{A}$ is. If $\bar{A}$ has complete consistency, a model can be established to estimate the weight of the reciprocal judgment matrix of interval number based on the idea of making the sum of the deviation value smallest.

$$\min \sum_{i,j} ap_{ij} + ad_{ij} \tag{6.44}$$

$$\text{s.t.} \begin{cases} a_{ij}^{\ L} wa_j \leq wa_i + ap_{ij}, i, j = 1, \ldots, n, i \neq j & (6.45) \\ wa_i \leq a_{ij}^{\ U} wa_j + ad_{ij}, i, j = 1, \ldots, n, i \neq j & (6.46) \\ \sum_{i=1}^{n} wa_i = 1 & (6.47) \\ wa_i \geq 0, ap_{ij}, ad_{ij} \geq 0, i = 1, \ldots, n & (6.48) \end{cases}$$

For two kinds of uncertainty preference such as complementary judgment matrix of interval number and utility value expressed by interval number, based on Formula (6.49), a model of solving weight can be established through introducing the deviation value.

$$b_{ij}^{\ L} \leq \frac{wb_i}{wb_i + wb_j} \leq b_{ij}^{\ U}, e_i^{\ L} \leq we_i \leq e_i^{\ U} \tag{6.49}$$

$\bar{B} = (b_{ij}^{\ L}, b_{ij}^{\ U})_{n \times n}$ is the complementary judgment matrix of interval number given by decision makers, and wb_i is the derived weight.

$\bar{E}_i = (e_i^L, e_i^U)$ is the utility value expressed by interval number that decision makers used and we_i can be set as the preference weight. For preference ordering of interval number $\bar{F}_i = [f_i^L, f_i^U]$, Formula (6.50) can be used to change the interval order into the utility value expressed by interval number to make the information consistent.

$$
f_i^{L'} = \frac{\dfrac{n - f_i^U}{n - 1} + \varepsilon}{\dfrac{n}{2} + n\varepsilon}, f_i^{U'} = \frac{\dfrac{n - f_i^L}{n - 1} + \varepsilon}{\dfrac{n}{2} + n\varepsilon} \tag{6.50}
$$

In Formula (6.50), $\varepsilon > 0$ is a small positive number and $f_i^{L'}, f_i^{U'}$ is the utility value expressed by interval number after changing.

The four kinds of sorting models of uncertainty preference have the same sorting mechanism via the suggested approach. During the process of specific group decision making, the reliability of group aggregation is relatively high. Multiple uncertainty preference can be changed into utility value of weight 0–1 after solving the model of unified weight.

6.3.2.3 Classification of Large-Scale Group Decision Based on Gray Clustering Gray clustering is a method that can gather some indices of observed objects and classify them based on a gray incidence matrix. It can determine whether they have very close relationships among many factors. This chapter uses gray clustering to classify for the utility value that expresses the decision-making preference of large-scale group decisions. It turned large-scale, unobvious convergence structure groups into a subgroup that has a consistent view and is composed of a certain number of decision makers through a given threshold value. Suppose that there are n observational objects and m data features for every observed object. Then a sequence can be obtained as follows:

$$
\begin{cases}
X_1 = [x_1(1), x_1(2), \ldots, x_1(n)] \\
X_2 = [x_2(1), x_2(2), \ldots, x_2(n)] \\
\cdots \cdots \cdots \cdots \cdots \cdots \cdots \\
X_m = [x_m(1), x_m(2), \ldots, x_m(n)]
\end{cases} \tag{6.51}
$$

For $i \leq j$, $i, j = 1, 2, \ldots, m$ the absolute correlation degree ε_{ij} of X_i and X_j can be calculated. So an upper triangular matrix

$$A = \begin{bmatrix} \varepsilon_{11} & \varepsilon_{12} & \cdots & \varepsilon_{1m} \\ & \varepsilon_{22} & \cdots & \varepsilon_{2m} \\ & & \ddots & \vdots \\ & & & \varepsilon_{mm} \end{bmatrix}$$

can be obtained that has $\varepsilon_{ij} = 1$; $i = 1, 2, \ldots, m$. Matrix A is called a *characteristic variable*. Then take critical value $r \in [0, 1]$ and $r \geq 0.5$. When $\varepsilon_{ij} \geq r(i \neq j)$, X_i and X_j can be regarded as the same type. Moreover, when r approaches 1, the number of classifications is greater and when r is smaller, the number of classifications is fewer.

In fact, the group classification has to consider the organizations, the departments that they belong to, and the decision makers' intrinsic attributes. But different decision makers also can have different views in a certain situation even if they are in the same organization and department because of conflicts of interest. So in some decision-making situations, the cluster features shown by group decision judgment preference become the main index for classification.

The decision group was divided into some classes based on their judgment preference after gray clustering. Moreover, the judgment view of intraclass decision makers is roughly the same.

6.3.2.4 Meta syntheses of Decision Group Intraclass Preference For a classification k after gray clustering, x_{ij}^k, $j = 1, \ldots, n$ is the value of judgment preference given by decision maker j in classification k. Suppose that the comprehensive preference of the decision group in this classification is x_{gi}^k, $j = 1, \ldots, n$. The deviation of comprehensive preference of the decision group in this classification and decision makers' own preference is $d^k = \sum_{i,j} x_{gi}^k - x_{ij}^k$. d^k can be shown as $d^k = \sum_{i,j} \left| x_{gi}^k - x_{ij}^k \right|$ in order to eliminate the influence of positive and negative deviation. In general, the pursuit of group decision making is to achieve consistency in the maximal degree. There must be minimal bifurcation between the comprehensive preference of the decision group that is in the same classification and decision makers' own preference. Finally, model $P_{6.9}$, which can solve the comprehensive preference of the decision group, can be obtained. w_i, $w_i \geq 0$ is the importance weight of decision makers who are in the same classification.

$$P_{6.9}: s.t. \begin{cases} \min d^k = \sum_{i,j} \left| w_i (w_{gj}^k - x_{ij}^k) \right| \\ \sum_j x_{gj}^k = 1 \\ 0 \leq x_{gj}^k \leq 1 \end{cases} \tag{6.52}$$

Then model $P_{6.9}$ can be changed as follows in order to solve easily. Suppose that $u_{ij} = w_i(x_{gj}^k - x_{ij}^k)$ and u_{ij} is either positive or negative. Also suppose that $u_{ij}' = \begin{cases} u_{ij}, u_{ij} \geq 0 \\ 0, u_{ij} < 0 \end{cases}$, $u_{ij}'' = \begin{cases} 0, u_{ij} \geq 0 \\ -u_{ij}, u_{ij} < 0 \end{cases}$,

$u_{ij} = u_{ij}' - u_{ij}''$, $\left| u_{ij} \right| = \left| u_{ij}' - u_{ij}'' \right| = \left| u_{ij}' + u_{ij}'' \right|$.

Through this transformation, the problem can be turned into model $P_{6.10}$:

$$P_{6.10}: \begin{cases} \min d^k = \sum_{i,j} (u_{ij}' + u_{ij}'') \\ s.t. \ u_{ij}' - u_{ij}'' = w_i(x_{gj}^k - x_{ij}^k) \\ \sum_j x_{gj}^k = 1 \\ 0 \leq x_{gj}^k \leq 1, \ u_{ij}', \ u_{ij}'' \geq 0 \end{cases} \tag{6.53}$$

Theorem 6.5

There must be an optimal solution for model $P_{6.10}$. The proof is omitted.

So how to determine the expert weight is a problem in model $P_{6.10}$? Trial calculation is used in this chapter to solve the problem, and the concrete method is shown as follows:

Step 1: Suppose

$$w_i = \frac{1}{m_k}$$

in model $P_{6.10}$ and m_k is the number of decision makers in classification k. After solving $P_{6.10}$, $w_{gi}^{\ k}$ can be obtained.

Step 2: Set $d_i = \Sigma_j |w_{gj}^{\ k} - x_{ij}^{\ k}|$ to be the sum of deviation of decision maker i and comprehensive preference in classification k. In general, as d_i is smaller, the view of decision maker i and comprehensive preference in this classification is closer. Then, the weights of decision makers are greater.

Step 3: Substitute w_i into $P_{6.10}$ to calculate again, and comprehensive preference in this classification $x_{gj}^{\ k}$ can be obtained. Then comprehensive preference considered the different weights of decision makers in this classification.

Definition 6.15

$w_i = 1/d_i \big/ \Sigma 1/d_i$, $i = 1, \ldots, m_k$ is the weight of decision makers in the decision group.

After obtaining the comprehensive preference of this classification, calculate the degree of correlation of decision makers in this classification and comprehensive preference $x_{gj}^{\ k}$ through the method of gray correlation. The correlation degree of decision maker i and comprehensive preference in classification k is $\varsigma_{gi}^{\ k}$. So the greater $\varsigma_{gi}^{\ k}$ is, the closer the consensus degree of the decision maker and comprehensive preference is. As $\varsigma_{gi}^{\ k} = 1$, the decision makers and comprehensive preferences have the same view.

Definition 6.16

The degree of consensus opinion that was given by decision makers in classification is $\varsigma_g^{\ k} = \varsigma_{gi}^{\ k}/n$.

As the value of $\varsigma_g^{\ k}$ is greater, the opinion of decision makers is more centralized and vice versa. During the process of group decision making, when $\varsigma_g^{\ k}$ is small, the views of decision makers in this classification have to be coordinated to achieve consistency.

Definition 6.17

Suppose $\varsigma_{gg}^{k} = \max\{\varsigma_{gi}^{k}|i = 1, \ldots, n\}$ and call decision maker g the core decision maker in classification k. Then there is maximal degree of correlation between decision makers and comprehensive preference.

When ς_{g}^{k} is small, the decision makers in the classification need to coordinate to adjust their own preferences. During the process of coordination, decision maker g can be thought as the benchmark. Other decision makers can adjust their views with decision maker g and adjust their judgments. It is easy to coordinate to achieve the best effect because after gray clustering the number of decision makers is not very large. (If the number of decision makers is still large after gray clustering, cluster it again.) Finally, consistent judgment can be obtained in every classification. Comprehensive preference is not used as an adjusting benchmark in this chapter because comprehensive preference is a modification benchmark that increases the difficulty of communication. So this is the main reason that classification decision and management of large-scale group decision is proposed in this chapter.

Comprehensive preference and consensus degree can be obtained through the method above for every classification using gray clustering.

6.3.2.5 Metasynthesis of Decision Group Preference between Classifications Based on Gray Cluster   Suppose that comprehensive preference between classifications is x_{gj} for comprehensive preference of every classification. Similarly, the deviation of comprehensive preference between classifications and comprehensive preference in every classification should be minimal. So suppose that the weight of every classification is w_k and model $P_{6.11}$ can be obtained, which can solve comprehensive preference between classifications.

$$P_{6.11} : \text{s.t.} \begin{array}{l} \min d = \sum_{k,j} \left| w_k (x_{gj} - x_{gj}^{k}) \right| \\ \sum_{j} x_{gj}^{k} = 1 \\ 0 \leq x_{gj}^{k} \leq 1 \end{array} \tag{6.54}$$

The method of trial calculation is used to solve the weight w_k of every classification. First, suppose that $w_k = 1/m$ and m is the number of classifications. Comprehensive preference between classifications x_{gj} can be obtained in model $P_{6.11}$. Then determine the weight of classification k through the deviation formula $d^k = \Sigma_j |x_{gj} - x_{gj}^k|$ of comprehensive preference and $w_k = 1/d^k \big/ \Sigma 1/d^k$.

Moreover, the number of decision makers in every classification affects the weight of classification. In general, based on the majority rule of group decision making, the greater the number of decision makers in this classification, the more important status the classification in the group and vice versa. So the number of decision makers in the classification and the degree of consistency are proposed to adjust the weight of every classification in this chapter.

Definition 6.18

Suppose w_k'' to be the weight of classification k in the large-scale decision group.

$$w_k'' = \sqrt{w_k w_k'} \times \varsigma_g^{\ k} \qquad (6.55)$$

In Formula (6.55), w_k is the weight of every classification based on the gap of comprehensive preferences in and between classifications when it was used as the weight of trial calculation. w_k' is to adjust the weight of classification based on the number of decision makers in every classification.

$$w_k' = \frac{m_k}{n},$$

n is the total number of the decision makers in the decision group. mk is the number of the decision makers in classification k. When m_k is larger, the weight of classification should be larger too and vice versa. $\varsigma_g^{\ k}$ is the degree of consistency between classifications. When the views of decision makers are more concentrated, the weight of classification should be higher.

Calculate the comprehensive preferences between classifications based on model $P_{6.11}$ after obtaining the weight of every classification. Similarly, model $P_{6.11}$ can change in the way of $P_{6.9}$ changing to $P_{6.10}$ and solve it easily.

6.3.3 Example Analysis

A work unit needed to select a talented person and inspected four candidates' comprehensive capacities. The first step was appraising the records of four candidates serving a term. Twenty-five experts who used multiple uncertainty preference gave scores to the four candidates. The utility values of four candidates can be obtained as follows based on the unified sorting method of uncertainty preference. x_i, $i = 1, 2, \ldots 25$ represent the preference opinions of 25 experts:

$x1 = 0.2, 0.3, 0.1, 0.4; x2 = 0.1, 0.3, 0.2, 0.4; x3 = 0.3, 0.2, 0.2,$
$0.2; x4 = 0.4, 0.1, 0.3, 0.2; x5 = 0.3, 0.5, 0.1, 0.1; x6 = 0.2, 0.2,$
$0.1, 0.5; x7 = 0.3, 0.3, 0.1, 0.3; x8 = 0.1, 0.1, 0.3, 0.5; x9 = 0.1,$
$0.6, 0.2, 0.1; x10 = 0.4, 0.15, 0.15, 0.3; x11 = 0.15, 0.25, 0.3, 0.3;$
$x12 = 0.25, 0.2, 0.25, 0.3; x13 = 0.1, 0.25, 0.15, 0.5; x14 = 0.10,$
$0.1, 0.2, 0.6; x15 = 0.5, 0.2, 0.15, 0.15; x16 = 0.2, 0.15, 0.35,$
$0.3; x17 = 0.3, 0.3, 0.3, 0.1; x18 = 0.2, 0.2, 0.3, 0.3; x19 = 0.1,$
$0.2, 0.4, 0.3; x20 = 0.4, 0.2, 0.3, 0.1; x21 = 0.4, 0.4, 0.1, 0.1;$
$x22 = 0.35, 0.25, 0.1, 0.3; x23 = 0.5, 0.2, 0.05, 0.25; x24 = 0.25,$
$0.25, 0.15, 0.35; x25 = 0.1, 0.25, 0.25, 0.4.$

The problem is who is the best according to the judgments of 25 experts in group 1?

Step 1: Calculate the correlation coefficient of experts based on the method of gray correlation clustering.

Suppose that the critical value of clustering $r = 0.9$ and use the correlation coefficient between the judgments of experts. The 25 experts can be divided into five groups. Group 1 is $\{x1, x6, x12, x16, x18\}$, group 2 is $\{x2, x8, x9, x11, x13, x19, x25\}$, group 3 is $\{x3, x5, x7, x14, x17, x24\}$, group 4 is $\{x4, x10, x20, x21, x22\}$, and group 5 is $\{x15, x23\}$.

Step 2: Aggregate the opinion between the classifications for five experts in group 1.

The comprehensive preference of experts in group 1 is $X_g^1 = \{0.2, 0.2, 0.3, 0.3\}$ according to the preference aggregating model $P_{6.10}$.

Calculate the weight of experts in group 1 according to

$$d_k = \sum_i \left| X_g - X_i^k \right|$$

and

$$w_k = \frac{1}{d_k} \Big/ \sum \frac{1}{d_k}.$$

$d_1 = 0.6$, $d_2 = 0.4$, $d_3 = 0.1$, $d_4 = 0.02$, $d_5 = 0.1$, $w_1 = 0.0023$, $w_2 = 0.0349$, $w_3 = 0.1395$, $w_4 = 0.6976$, and $w_5 = 0.1395$ can be obtained.

Calculate the comprehensive preference of group 1 again using the weight of experts in group 1. Then the comprehensive preference of group 1 can be obtained as 0.2070, 0.1960, 0.2967, 0.3114.

Calculate the correlation coefficient of the experts' judgments and comprehensive preference of group 1 using the original judgments and comprehensive preference of experts in group 1, which is 0.82, 0.85, 0.89, 0.80, 0.80. The arithmetic mean is 0.83 and it is the consistent degree of the experts' views in group 1.

In the same way, the comprehensive preference of group 2 is 0.1052, 0.2569, 0.2467, 0.3911 and the consistency degree is 0.97. The comprehensive preference of group 3 is 0.2846, 0.2662, 0.2046, 0.2277 and the consistency degree is 0.95. The comprehensive preference of group 4 is 0.3875, 0.2167, 0.1792, 0.2169 and the consistency degree is 0.97. The comprehensive preference of group 5 is 0.5, 0.2, 0.1333, 0.1667 and the consistency degree is 0.99. Table 6.2 shows, the results. Calculate the comprehensive judgments and concentration ratio of every group according to the method above.

The consistency degree of group 1 is low, and other groups' views are consistent from Table 6.2.

Step 3: Aggregate the opinions between groups.

Calculate the aggregated opinions between groups according to the method as follows for the experts of group 5.

Table 6.2 The Aggregation Degree of Comprehensive Judgments and Views of Group

GROUP	X1	X2	X3	X4	CONSISTENCY DEGREE
Group1	0.2066	0.2018	0.2978	0.3223	0.83
Group2	0.1052	0.2569	0.2467	0.3911	0.97
Group3	0.2846	0.2662	0.2046	0.2277	0.95
Group4	0.3875	0.2167	0.1792	0.2169	0.97
Group5	0.5	0.2	0.1333	0.1667	0.99

a. According to

$$\min = \sum_{i,k} \left| X_g - X_i^k \right|$$

$$\text{s.t. } \sum_g X_g = 1$$

the comprehensive preference of group 5 is 0.2846, 0.2167, 0.2467, 0.252.

b. Calculate the weight of every group's opinion according to $d_k = \Sigma_i |X_g - X_i^k|$, $w_i = 1/d_i \big/ \Sigma 1/d_i$. $d_1 = 0.2143$, $d_2 = 0.3587$, $d_3 = 0.1159$, $d_4 = 0.2055$, $d_5 = 0.4308$, $w_1 = 0.2005$, $w_2 = 0.1198$, $w_3 = 0.3708$, $w_4 = 0.2091$, and $w_5 = 0.0998$ can be obtained.

Adjust the weight of every group based on the number of each group's experts. $w_1' = 5/25 = 0.2$, $w_2' = 7/25 = 0.28$, $w_3' = 6/25 = 0.24$, $w_4' = 5/25 = 0.2$, $w_5' = 2/25 = 0.08$.

According to the ultimate weight of each group $w_i'' = \sqrt{w_i \times w_i'} \times \alpha$, the weights of the five groups are 0.1818, 0.1944, 0.31, 0.2170, 0.0968 after normalizing. Then calculate the comprehensive preference again and it is 0.2787, 0.2355, 0.2173, 0.2684. So the order of the four candidates is 1, 3, 4, 2.

6.4 Aggregation of Two Kinds of Preference Information Considering Consistency Coordination

6.4.1 Background

In the process of group decision making, all decision problems are based on the preferences of decision makers, so how to aggregate individual preferences of multiple decision makers into group preference becomes the core of group decision making. Current research focuses on aggregation models of preferences, measuring methods of group consistency in the aggregation process [8, 38, 40]. From the existing literature [32, 34, 35, 48], the following three points can be seen:

1. The consistency of preference information has been considered in the group aggregation process in the literature [8, 38, 40], but the coordination between individual consistency and group consistency has not been taken into account. Only a few literature works emphasize adjusting individual or group consistency singly.
2. The method of how to simultaneously improve individual and group consistency is lacking. In many cases, there are not only similarities but also differences between individual consistency and group consistency.
3. There is little research on group fixed rights in the existing preference aggregation method. In fact, though all group members take the decision-making task, they may be very different in judging level from others.

Therefore, this chapter proposes a consistency adjustment method considering both individual consistency and group consistency based on the reciprocal judgment matrices and complementary judgment matrices given by decision makers. The adjustment will not be limited to an adjustment of a single matrix item but rather focus on overall items of the judgment matrix. Then a model is built for decision makers' weights that considers individual and group consistency indices. So that group preference should be possibly similar to all individual preferences, an aggregation model of preferences is established based on weighting of decision makers.

6.4.2 Consistency Modification Model for Group Decision Making

Definition 6.19

Let $A = (a_{ij})$ be an $n \times n$ judgment matrix, $\Omega = \{1, 2, \ldots, n\}$, if $a_{ij} \geq 0$, $a_{ij} = 1/a_{ji}$ and $a_{ii} = 1$, for each $i, j, \in \Omega$, then $A = (a_{ij})$ is a reciprocal judgment matrix.

Definition 6.20

Let $A = (a_{ij})$ be an $n \times n$ judgment matrix, $\Omega = \{1, 2, \ldots, n\}$, if $1 \geq a_{ij} \geq 0$, $a_{ij} = 1 - a_{ji}$, and $a_{ii} = 0.5$, for each $i, j, \in \Omega$, then $A = (a_{ij})$ is a complementary judgment matrix.

6.4.2.1 Consistency Check Method for Individual For reciprocal judgment matrix A of order n, an individual consistency measure was suggested, called the *consistency ratio*, given by

$$CR = \frac{CI}{RI(n)} \tag{6.56}$$

where CI is the consistency index given by $CI = (\lambda_{\max} - n)/(n - 1)$, $RI(n)$ is the random consistency index for matrices of order n, and $\lambda_{\max}$ is the principal eigenvalue of the judgment matrix. Judgment matrix A is of acceptable individual consistency if $CR < 0.1$.

For complementary judgment matrix A of order n, an individual consistency measure was suggested, given by

$$FCI^* = \frac{2}{n(n-1)} \sum_{i=1}^{n-1} \sum_{j=i+1}^{n} (a_{ij} - w_{ij})^2 \tag{6.57}$$

where $w = (w_1, w_2, \ldots, w_n)$ is the priority vector of judgment matrix A, w_{ij} is given by

$$w_{ij} = \frac{1}{2}(w_i - w_j) + 0.5,$$

for each $i, j, \in, \Omega = \{1, 2, \ldots, n\}$. Judgment matrix A is of acceptable individual consistency if $FCI^* < 0.04$.

Taking into account the difference in individual consistency definitions of two types of judgment matrices, the relative deviation index of individual consistencies is defined in order to uniformly compare individual consistencies of all judgment matrices given by decision makers.

Definition 6.21

Let RD be a relative deviation of reciprocal or complementary judgment matrix individual consistency index value and its threshold value (0.1 or 0.04) if RD is given by

$$RD = \frac{0.1 - CR(k)}{0.1}$$

in the reciprocal judgment matrix and by

$$RD = \frac{0.04 - FCI^*}{0.04}$$

in the complementary judgment matrix, then RD is the relative deviation index of individual consistencies. Judgment matrix A is of acceptable individual consistency if $RD > 0$.

RD is equivalent to individual consistency indices non-dimension of the reciprocal and complementary judgment matrix. Thus, the individual consistency indices of the reciprocal and complementary judgment matrix would be comparable. Larger RD means greater individual consistency.

6.4.2.2 Consistency Check for Group   In addition to the individual consistency check, this chapter carries out a group consistency check based on gray correlation. According to the definition of gray correlation, the degree of relevancy of judging information given by decision makers can be reflected through calculating the correlation (considered not only the absolute value of information but also the direction of information). The greater the correlation is, the more similar the preferences given by decision makers are and the closer those are to the group preference. Thus, checking group consistency of each matrix by gray correlation can not only contribute to consistency analysis on the preference information but also provide the basis for aggregation of all preference information.

Definition 6.22

Let A_k be a judgment matrix given by decision maker k, $k = 1, 2, \ldots, s$, if the priority vector of judgment matrix A_m is of acceptable individual consistency and the closest to the average priority vector of all judgment matrices, for $m \in \Omega$, $1 \leq m \leq s$, then decision maker m is the core decision maker.

According to the correlation measure, the core decision maker should be selected first, and the priority vector given by him will be the reference sequence. Then the correlation of other priority vectors will be calculated to the reference sequence. If a correlation value is less than a specific value, the corresponding judgment is not of acceptable group consistency. The specific process is as follows:

1. The average vector calculation making use of weighted arithmetic average method: Let $W^g = \{w_i^g | i = 1, 2, \ldots, n\}$ be the average vector, given by

$$w_i^g = \frac{1}{s} \sum_{k=1}^{s} w_i^k, \ i = 1, 2, \ldots, n \qquad (6.58)$$

where w_i^k is the weight value given by judgment matrix A_k.

2. Selection of reference sequence denoted W^m:

$$\min_{1 \leq k \leq s} \left\{ \sum_{i=1}^{n} (w_i^k - w_i^g)^2 \right\} \qquad (6.59)$$

Suppose that when

$$\sum_{i=1}^{n} (w_i^k - w_i^g)^2$$

obtains the minimum value, k equals m. At the same time, if judgment matrix A_m is of acceptable individual consistency, decision maker m is the core decision maker. Otherwise, seek out the decision maker in others when

$$\sum_{i=1}^{n} (w_i^k - w_i^g)^2$$

is the lowest and check the corresponding individual consistency, and then repeat this step until the election of the core decision maker. The priority vector given by the core decision maker m is the reference sequence denoted W^m.

3. Correlation measure based on the reference sequence

$$\gamma^k = \frac{1}{n} \sum_{i=1}^{n} \frac{m + \xi M}{\Delta_k(i) + \xi M}, \quad \xi \in (0, 1) \tag{6.60}$$

where $\Delta_k(i)$ is given by $\Delta_k(i) = |w_i^k - w_i^m| (i = 1, 2, \ldots, n)$, M is the maximal value given by $M = \max_k \max_i \Delta_k(i)$, m is the minimal value given by $m = \min_k \min_i \Delta_k(i)$. The correlation of judgment matrix A_k is denoted γ^k.

Definition 6.23
Let T be the average of all correlation values, given by

$$T = \frac{1}{s} \sum_{k=1}^{s} \gamma^k.$$

If γ^k is larger than T, judgment matrix A_k is of acceptable group consistency. Otherwise, judgment matrix A_k needs to be adjusted. Then T is the group consistency index threshold.

6.4.2.3 Modification Method Based on Group and Individual In order to obtain more scientific results from group decision making, it is important to analyze the differences between group members and coordinate some judgment matrices with others. In this chapter, after the individual and group consistency check, judgment matrices that are of unacceptable consistency are coordinated to meet consistency requirements. For the complementary judgment matrix A_k needing modification, each item denoted a_{ij} (except items in which i equals j) plus a margin denoted x_{ij} makes a new item denoted a'_{ij}, which should ensure that modified judgment matrix A_k is not only of acceptable consistency but also close to the original matrix. The modified complementary judgment matrix still satisfies Formulas (6.61) and (6.62), so Formulas (6.63) and (6.64) can be found:

$$a'_{ij} + a'_{ji} = 1 \qquad\qquad (6.61)$$

$$0 \le a'_{ij} \le 1 \qquad\qquad (6.62)$$

$$x_{ji} = -x_{ij} \qquad\qquad (6.63)$$

$$-a_{ij} \le x_{ij} \le 1 - a_{ij} \qquad\qquad (6.64)$$

According to Formulas (6.60) and (6.64), the following nonlinear programming model can be established:

$$\min z = \sum_{i=1}^{n-1} \sum_{j=i+1}^{n} x_{ij}^2$$

$$\begin{cases} a'_{ij} = a_{ij} + x_{ij}(i \ne j) \\ FCI^* = \dfrac{2}{n(n-1)} \displaystyle\sum_{i=1}^{n-1} \sum_{j=i+1}^{n} (a'_{ij} - w_{ij})^2 \\ \dfrac{1}{n} \displaystyle\sum_{i=1}^{n} \dfrac{m + \xi M}{\Delta_k(i) + \xi M} \ge T \\ FCI^* < 0.04 \\ a'_{ij} + a'_{ji} = 1 \\ -a_{ij} \le x_{ij} \le 1 - a_{ij} \end{cases} \qquad\qquad \text{P}_{6.12}$$

For the reciprocal judgment matrix, the numerical relationship between x_{ij} and x_{ji} is more complicated and it is difficult to establish a simple and effective correction model. Therefore, at first translate the reciprocal judgment matrix A_k needing modification into complementary judgment matrix B_k, making use of Formula (6.65) (here let a equal 100), then modify complementary judgment matrix B_k. Through model $\text{P}_{6.12}$, the original reciprocal or complementary judgment matrices of unacceptable individual or group consistencies will become complementary judgment matrices of acceptable individual and group consistencies, possessing new values of RD and $\gamma^k (k \in \{1, 2, \ldots, s\})$.

$$b_{ij} = \log_a a_{ij} + 0.5 \tag{6.65}$$

6.4.3 Weight Definition of Decision Maker Based on Consistency of Group and Individual

The value of the individual consistency index of the judgment matrix directly reflects the logical degree of judging information given by decision makers, which is equal to the credibility of the information. In the group consistency check process, it can be seen that the view of the core decision maker plays a critical and decisive role and largely reflects the group preference. That is, if one decision maker's preference has greater correlation, it will be more able to reflect group preference. Therefore, considering the individual and group consistencies, a weight definition model of decision makers is built based on the relative deviation of individual consistency denoted RD and correlation of judgment matrix denoted γ_k:

$$P^k = \beta \frac{RD^k}{\sum\limits_{i=1}^{s} RD^i} + (1-\beta) \frac{\gamma^k}{\sum\limits_{i=1}^{s} \gamma^i} \tag{6.66}$$

where β is set as the weights of individual consistency and group consistency, and its value is between 0 and 1. The greater the value is, the more attention is paid to the individual consistency.

Weights of decision makers denoted P^k can be found through Formula (6.66). So that group preference should be possibly similar to all individual preferences, the following aggregation model of preferences is established:

$$\min Z = \sum_{k=1}^{s} \sum_{i=1}^{n} P^k (w_i - w_i^k)^2$$

$$\begin{cases} \sum\limits_{i=1}^{n} w_i = 1 \\ w_i > 0, \, i = 1, 2, \ldots, n \end{cases} \qquad \text{P}_{6.13}$$

After the solution of model $P_{6.13}$, the optimal solution can be obtained, which is also the weight vector of group aggregated preference, denoted $W^* = \{w_i | i = 1, 2, \ldots, n\}$.

Theorem 6.6

The optimal solution of model $P_{6.13}$ is $W^* = (\sum_{k=1}^{s} P^k w_1^k, \sum_{k=1}^{s} P^k w_2^k, \cdots, \sum_{k=1}^{s} P^k w_n^k)$.

Proof

According to the Lagrange multiplier method, construct the following function: $f(w_1, w_2, \ldots, w_n, \lambda) = \sum_{k=1}^{s} \sum_{i=1}^{n} P^k (w_i - w_i^k)^2 + \lambda(\sum_{i=1}^{n} w_i - 1)$.

So the optimal solution of model $P_{6.13}$ equals the minimum value of the function above.

Then seek the first-order partial derivatives as follows:

$$\frac{\partial f}{\partial w_i} = 2 \sum_{k=1}^{s} P^k (w_i - w_i^k) + \lambda, \quad \frac{\partial f}{\partial \lambda} = \sum_{i=1}^{n} w_i - 1$$

Let

$$\frac{\partial f}{\partial w_i} = 0$$

and

$$\frac{\partial f}{\partial \lambda} = 0.$$

It can be obtained that:

$$\begin{cases} w_i = \sum_{k=1}^{s} P^k w_i^k, i = 1, 2, \ldots, n \\ \lambda = 0 \end{cases}$$

Hence, the only extreme value of the function $f(w_1, w_2, \ldots, w_n, \lambda)$ has been obtained.

Because

$$\frac{\partial^2 f}{\partial w_i^2} \geq 0$$

and

$$\frac{\partial^2 f}{\partial \lambda^2} \geq 0,$$

the only extreme value is the minimum.

It should be noted that model $P_{6.13}$ is the same with the case of the weights denoted $w_i(i = 1, 2, \ldots, n)$ without more constraints. If some weight denoted $w_X(1 \leq X \leq n)$ should satisfy the constraint that $w_X \in (w_X^L, w_X^U)$, model $P_{6.13}$ can be extended into the following:

$$\min Z = \sum_{k=1}^{s} \sum_{i=1}^{n} P^k (w_i - w_i^k)^2$$

$$\begin{cases} \sum_{i=1}^{n} w_i = 1 \\ w_i > 0, i = 1, 2, \cdots, n, i \neq X \\ w_X^L < w_X < w_X^U, X \in [1, n] \end{cases} \qquad P_{6.14}$$

The meaning of parameters in model $P_{6.14}$ is the same as that in model $P_{6.13}$.

6.4.4 Example Analysis

Earned value management (EVM) is an effective management tool for cost and scheduling. Establishing an EVM performance evaluation system can measure project progress and cost scientifically, in order to track and predict a project's performance, which is difficult in weighting of performance evaluation indices. As a result, the key points of progress management of the project are difficult to determine. The main difficulty lies in no consensus as a result of knowledge structure differences, cognitive structure differences, and largely different views of evaluation experts, of which the points cannot be grasped easily. Therefore, this chapter attempts to determine a consistent and coordinated approach for groups, focusing the concerns of experts in order to identify differences and implement effective improvement.

The EVM performance evaluation system consists of five indicators: cost performance index, budget after completion, expectations after completion, schedule performance index, and variance after completion. For these five indicators, seven experts give the following seven judgment matrices to rate their importance:

$$
A_1 = \begin{bmatrix} 1 & 2 & 3 & 1 & 1/2 \\ 1/2 & 1 & 2 & 1/2 & 1/3 \\ 1/3 & 1/2 & 1 & 1/2 & 1/4 \\ 1 & 2 & 2 & 1 & 1/2 \\ 2 & 3 & 4 & 2 & 1 \end{bmatrix} \quad
A_2 = \begin{bmatrix} 1 & 2 & 2 & 3/2 & 1/2 \\ 1/2 & 1 & 2/3 & 1/2 & 1/2 \\ 1/2 & 3/2 & 1 & 1/2 & 1/5 \\ 2/3 & 2 & 2 & 1 & 1/2 \\ 2 & 2 & 5 & 2 & 1 \end{bmatrix}
$$

$$
A_3 = \begin{bmatrix} 1 & 2 & 4 & 3/2 & 1 \\ 1/2 & 1 & 2 & 1/2 & 1 \\ 1/4 & 1/2 & 1 & 1/2 & 1/3 \\ 2/3 & 2 & 2 & 1 & 1/2 \\ 1 & 1 & 3 & 2 & 1 \end{bmatrix},
$$

$$
A_4 = \begin{bmatrix} 0.5 & 0.7 & 0.8 & 0.6 & 0.4 \\ 0.3 & 0.5 & 0.6 & 0.4 & 0.2 \\ 0.2 & 0.4 & 0.5 & 0.4 & 0.1 \\ 0.4 & 0.6 & 0.6 & 0.5 & 0.4 \\ 0.6 & 0.8 & 0.9 & 0.6 & 0.5 \end{bmatrix}, \quad
A_5 = \begin{bmatrix} 0.5 & 0.7 & 0.8 & 0.5 & 0.3 \\ 0.3 & 0.5 & 0.7 & 0.4 & 0.2 \\ 0.2 & 0.3 & 0.5 & 0.3 & 0.1 \\ 0.5 & 0.6 & 0.7 & 0.5 & 0.4 \\ 0.7 & 0.8 & 0.9 & 0.6 & 0.5 \end{bmatrix}
$$

$$
A_6 = \begin{bmatrix} 0.5 & 0.6 & 0.9 & 0.5 & 0.3 \\ 0.4 & 0.5 & 0.6 & 0.3 & 0.2 \\ 0.1 & 0.4 & 0.5 & 0.3 & 0.1 \\ 0.5 & 0.7 & 0.7 & 0.5 & 0.4 \\ 0.7 & 0.8 & 0.9 & 0.6 & 0.5 \end{bmatrix},
$$

$$
A_7 = \begin{bmatrix} 1 & 2 & 1/2 & 2 & 1 \\ 1/2 & 1 & 3/2 & 1 & 1/4 \\ 2 & 2/3 & 1 & 1/3 & 1/5 \\ 1/2 & 1 & 3 & 1 & 1/4 \\ 1 & 4 & 5 & 4 & 5 \end{bmatrix}.
$$

1. Consistency check method for individual:

Because $W^1 = [0.2171, 0.1228, 0.0812, 0.2011, 0.3778]$ and λ_{max} equals 5.0488, CR equals 0.011, less than 0.1; thus, the judgment matrix A_1 is of acceptable individual consistency.

Because $W^2 = [0.2180, 0.1114, 0.1072, 0.1859, 0.3775]$ and λ_{max} equals 5.1491, CR equals 0.03, less than 0.1; thus, the judgment matrix A_2 is of acceptable individual consistency.

Because $W^3 = [0.2878, 0.1779, 0.0806, 0.1961, 0.2576]$ and λ_{max} equals 5.2726, CR equals 0.061, less than 0.1; thus, the judgment matrix A_3 is of acceptable individual consistency.

Because $W^4 = [0.2312, 0.1688, 0.1437, 0.2, 0.2563]$, FCI^* equals 0.032, less than 0.04; thus, the judgment matrix A_4 is of acceptable individual consistency.

Because $W^5 = [0.2188, 0.175, 0.1312, 0.2125, 0.2625]$, FCI^* equals 0.037, less than 0.04; thus, the judgment matrix A_5 is of acceptable individual consistency.

Because $W^6 = [0.2187, 0.1688, 0.1312, 0.2188, 0.2625]$, FCI^* equals 0.041, larger than 0.04; thus, judgment matrix A_6 is of unacceptable individual consistency.

Because $W^7 = [0.2075, 0.1145, 0.13, 0.1483, 0.3997]$ and λ_{max} equals 5.773, CR equals 0.17, larger than 0.1; thus, the judgment matrix A_7 is of unacceptable individual consistency.

It can be concluded that the individual consistencies of judging information given by seven experts are mostly great, except expert 6 and expert 7 in poor consistency levels.

2. Consistency check method for group:

By substitution of the weight vector values above into Formulas (6.58) and (6.59), it can be gained that m equals 5 and judgment matrix A_5 is of acceptable individual consistency. Thus, the weight vector of judgment matrix A_5 is the reference sequence. According to Formula (6.60) (take ξM equal to 0.37), correlation values of other weight vectors are obtained as follows: $\gamma^1 = 0.71$, $\gamma^2 = 0.70$, $\gamma^3 = 0.66$, $\gamma^4 = 0.87$, $\gamma^6 = 0.97$, $\gamma^7 = 0.69$.

Hence, the group consistency index threshold denoted T equals 0.7 given by

$$T = \frac{1}{7}\sum_{k=1}^{7}\gamma^{k}.$$

3. Consistent difference matrix:

To more clearly reflect the individual and group consistency level of each decision maker, a consistent difference matrix is used that divides all consistency levels into three kinds such as high individual consistency, high group consistency, and low individual and group consistency (Figure 6.2).

Where the abscissa is RD, by Definition 6.21, larger RD means greater individual consistency, and the judgment matrix is of unacceptable individual consistency if $RD < 0$. The vertical axis is γ^{k}, and larger γ^{k} means greater group consistency. As T equals 0.7, the judgment matrix is of unacceptable group consistency if $\gamma^{k} < 0.7$.

From the consistent difference matrix above, it can be seen that judgment matrices A_1, A_2, A_4, and A_5 are of acceptable individual and group consistency; judgment matrix A_3 is of unacceptable group consistency; judgment matrix A_6 is of unacceptable individual consistency; and judgment matrix A_7 is of unacceptable individual and group consistency. Therefore, judgment matrices A_3, A_6, and A_7 need to be modified.

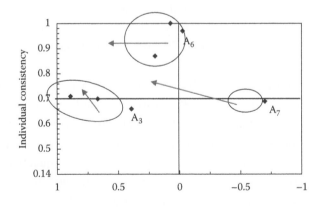

Figure 6.2   Individual and group consistency.

4. Consistent modification:

After modification, judgment matrices A_3, A_6, and A_7 become consistent as follows:

$$A_3 = \begin{bmatrix} 0.5 & 0.6605 & 0.802 & 0.588 & 0.5 \\ 0.3395 & 0.5 & 0.6705 & 0.3495 & 0.5 \\ 0.198 & 0.3295 & 0.5 & 0.3495 & 0.2612 \\ 0.412 & 0.6505 & 0.6505 & 0.5 & 0.3495 \\ 0.5 & 0.5 & 0.7388 & 0.6505 & 0.5 \end{bmatrix};$$

$$A_6 = \begin{bmatrix} 0.5 & 0.6037 & 0.886 & 0.5 & 0.3 \\ 0.3963 & 0.5 & 0.61 & 0.3 & 0.2 \\ 0.114 & 0.39 & 0.5 & 0.3 & 0.13 \\ 0.5 & 0.7 & 0.7 & 0.5 & 0.4 \\ 0.7 & 0.8 & 0.87 & 0.6 & 0.5 \end{bmatrix};$$

$$A_7 = \begin{bmatrix} 0.5 & 0.6515 & 0.3495 & 0.6505 & 0.5 \\ 0.3485 & 0.5 & 0.6 & 0.5 & 0.2 \\ 0.6505 & 0.4 & 0.5 & 0.2612 & 0.1505 \\ 0.3495 & 0.5 & 0.7388 & 0.5 & 0.2 \\ 0.5 & 0.8 & 0.8495 & 0.8 & 0.5 \end{bmatrix}.$$

The new weight vector and consistency values of judgment matrices A_3, A_6, and A_7 are as follows:

$W^3 = [0.2345, 0.1919, 0.1455, 0.2037, 0.2244]$, $W^6 = [0.2183, 0.169, 0.133, 0.2188, 0.2609]$, $FCI^*(3) = 0.021 < 0.04$, $\gamma^3 = 0.86 > 0.7$; $FCI^*(6) = 0.038 < 0.04$, $\gamma^6 = 0.96 > 0.7$
$W^7 = [0.2095, 0.178, 0.1644, 0.1868, 0.2593]$,
$FCI^*(7) = 0.036 < 0.04$, $\gamma^7 = 0.75 > 0.7$.

It can be seen that after modification, judgment matrices A_3, A_6, and A_7 are of acceptable individual and group consistency.

5. Weight definition of decision makers and aggregation of preferences:

Let β equal to 0.2. (Because individual consistency levels are good, group consistencies are paid more attention to.)

According to Formula (6.66), the weights of decision makers decision levels are differentiated to some extent as follows: $P^1 = 0.17$, $P^2 = 0.15$, $P^3 = 0.16$, $P^4 = 0.13$, $P^5 = 0.14$, $P^6 = 0.14$, $P^7 = 0.11$.

By substitution of P^k and W^k into model $_{6.12}$, the optimal solution can be gained as follows: $W^* = (0.2214, 0.1578, 0.1272, 0.2015, 0.2921)$.

Therefore, the weights of the five evaluation indicators—content of cost performance index, budget after completion, expectations after completion, schedule performance index, and variance after completion—are 0.2214, 0.1578, 0.1272, 0.2015, and 0.2921. Integrating the value results of performance indices, a project's performance can be tracked and evaluated to conform the scope, schedule, and cost of projects.

6.5   Summary and Future Research

The weight method of fuzzy linear programming is proposed on the hybrid uncertain comparison matrix. The index of degree of consistency of the hybrid uncertain comparison matrix is defined. The two-stage method for the weight is established to solve the problem for which there are many optimal solutions. Then, the weight possible distribution can be estimated. For especially sensitive problems, particularly those involving their own interest or confidential issues, decision makers may not like to express their preference views, which leads to the incomplete element in the hybrid uncertain comparison matrix. In the future this problem should be studied.

We researched four kinds of uncertain preference information aggregation approaches and proposed a new method for determining the experts' weights in group decision making that is based on a lot of structural uncertain preference information. The model is clear and simple to use. It has considerable reference value for the weight solving of relative preference information to introduce the deviation variables to solve interval number complementary judgment matrices, interval number utility values, and interval number preferences. The next step of research is to aggregate more kinds of uncertain preference information and develop a practical group decision-making support system in the environment of a web based on specific algorithms.

Due to the development of a democratic decision-making process, more decision makers participate in the decision. A new research goal is how to coordinate the bifurcation of the decision group and achieve a consistent decision in the situation of many decision makers. This chapter uses the method of gray clustering to aggregate the no obvious character group to the group that has the similar view. Through gray clustering, the coordination problem of large-scale decision makers' views can be changed into small scale and small scale coordination can decrease the difficulty of decision making. This chapter proposed the method of selection of the core decision maker. By determining the core decision maker, the communication among groups can increase despite bifurcation in groups. The method mentioned in this chapter is very simple and has practical and theoretical value.

Individual and group consistency check methods of the reciprocal or complementary judgment matrices are proposed in which modification models are built based on improving the consistency levels of judgment matrices. Then weights of decision makers are determined and an aggregation model is proposed that provides a new way to aggregate various types of preference information. The models are applied to determine the weights of EVM performance evaluation indices. This chapter focuses on two types of preference, and future research will focus on the measure and improvement of individual and group consistency under uncertain conditions. In addition, the consistency of more types of preferences will be the focus of attention.

This chapter studies aggregation methods for many types of preference information. In fact, aggregation of many types of preferences is of great significance while having greater value in the actual decision-making process. But in this context, the following three difficult issues have yet to be addressed:

1. Internal consistency mechanisms of various preferences: In the consistent processing of many types of information, an assumption must be proposed that there is a uniform judging criteria of information. But no effective method has been found to check the assumption's existence.

2. The existing consistent methods transform several kinds of information through a number of ways, but this transformation has not been strongly argued in the case when a simple transitive preference does not attest to the effectiveness of information transformation.

3. In aggregation of many types of information, it has not been determined whether the preference type of decision makers matters to the aggregation result or not. In fact, reasonable and effective forms of preference will certainly be beneficial to the group decision-making process; however, whether that will be related to the results or not is still unknown.

7

AGGREGATION METHOD OF MULTIPLE STAGES FOR GROUP DECISION MAKING

7.1 Introduction

Because of the differences in decision makers' life experiences, work experiences, psychological qualities, judgment levels, external environments, individual preferences, and others factors, different decision makers will prefer different structures, for example, judgment matrix, utility value, preference sequence value, linguistic value, etc. As objectives change, the decision makers' understanding of the objective things ranges from the shallower to the deeper. In addition, decision makers should research multiple stages of the problem in order to make a comprehensive assessment; this is because the stages influence each other and the preference form of the decision maker may be different in each stage. So, how to aggregate the preference information in multiple stages and structure forms is worth studying.

7.2 Aggregation Method of Reciprocal and Complementary Comparison Matrix with Multiple Stages

Academic circles have paid attention to certain preference aggregations in different preference structures, but there is no literature regarding the preference information aggregation in the multiple stage and multiple structure form. Several problems should be solved: confirm the weight of the decision group and stage reasonably, survey the problem dynamically, and make a scientific decision. The decision makers' preference belongs to subjective judgment and is different with the objective attribute data. There exists the problem of whether the decision makers' logic is consistent in the aggregated process of multiple-stage and multiple-judge preference. So, this section

177

researches the aggregating method of the reciprocal judgment matrix and complementary judgment matrix.

7.2.1 Transform for Two Identical Types of Preference

First transform the complementary judgment matrix into a reciprocal judgment matrix according to Formula (7.1). b_{ij} is the decision maker's judgment value based on comparing alternative i and alternative j.

$$a_{ij} = \frac{b_{ij}}{b_{ji}}, \, i, j = 1, \dots, n \tag{7.1}$$

The two-structure form is transformed into one according to Formula (7.1), denoted as A_i^k, $i = 1, \dots, t, k = 1, \dots, m$ is the judgment matrix of decision maker k in stage i. Regarding A_i^k, first calculate its export weight w_{ij}^k and consistency ratio CR_i^k according to the characteristic root method; $j = 1, \dots, n$ is the weight value and consistency level of alternative j given by decision maker k in stage i.

7.2.2 Weight Model of the Decision Maker and Rank of the Alternative

Many literature works have confirmed the weight of the decision maker according to excavating the preference information of the decision maker; this is a new exploration. Generally, the weight of the decision maker should be decided by synthesizing his or her political, ideological, moral, professional, psychological, and physical qualities; in some conditions it is not exact to confirm the decision maker's weight according to one judgment. But the decision maker in a multiple-stage process through multiple judgment with the weight based on this condition is more exact.

1. **Confirm the weight of decision makers based on consistency level.** It has been put forward that good decisions are mostly based on consistent judgment, and the consistency reflects coordination of logical relation in [131]. It also concludes the following by analogy: the dependability will improve when the decision makers have no prejudice and $CR \le 0.1$. So this section confirms the weight of decision

makers based on the difference of CR and calculates the average consistency level based on CR_i^k according to Formula (7.2), denoted as CR_i^k.

$$CR^k = \sqrt[t]{\prod_{i=1}^{t} CR_i^k}, \ k = 1, \dots, m. \tag{7.2}$$

Definition 7.1
Take

$$e_{CR}^k = \frac{\dfrac{\min\limits_{k=1,\dots,m} CR^k}{CR^k}}{\sum\limits_{k=1}^{m} \dfrac{\min\limits_{k=1,\dots,m} CR^k}{CR^k}}, \ k = 1, \dots, m$$

as the weight of decision maker k based on his own judgement consistency.

Based on this definition in t stages, the smaller the average consistency ratio of decision makers, the greater e_{CR}^k is.

2. **Confirm the weight of decision makers based on the deviation of decisions.** Suppose that the decision makers have the same weight in stage i. So the group preference w_{ij}, $i = 1, \dots, t, j = 1, \dots, n$ of this stage has the following feature: the deviation between the group preference and individual preference is the smallest; therefore, so the following formula is right if one uses $\sum_{j=1}(w_{ij} - w_{ij}^k)^2$ to measure the deviation between the group and decision maker's judgments.

$$\min Q = \sum_{j,k} (w_{ij} - w_{ij}^k)^2, \ i = 1, \dots, t$$

$$\mathrm{P}_{7.1} \begin{cases} \displaystyle\sum_{j=1}^{n} w_{ij} = 1, \ i = 1, \dots, t \\[2mm] w_{ij} \geq 0, \ i = 1, \dots, t, j = 1, \dots, n \end{cases}$$

Theorem 7.1

Denote the optimal solution of $P_{7.1}$ as w_{ij}^{*} and

$$w_{ij}^{*} = \frac{\displaystyle\sum_{k=1}^{m} w_{ij}^{k}}{m}.$$

Proof

Suppose that $Q' = \sum_{j,k}(w_{ij} - w_{ij}^{k})^{2} - \lambda(1 - \sum_{j=1}^{n} w_{ij})$,

so

$$\begin{cases} \dfrac{\partial Q'}{\partial w_{i1}} = 2\displaystyle\sum_{k}(w_{i1} - w_{i1}^{k}) - \lambda, \ i = 1, \ldots, t \\[2mm] \ldots \\[2mm] \dfrac{\partial Q'}{\partial w_{in}} = 2\displaystyle\sum_{k}(w_{in} - w_{in}^{k}) - \lambda, \ i = 1, \ldots, t \\[2mm] \dfrac{\partial Q'}{\partial \lambda} = 1 - \displaystyle\sum_{j=1}^{n} w_{ij} \end{cases}$$

Then let

$$\begin{cases} \dfrac{\partial Q'}{\partial w_{i1}} = 0 \\[2mm] \ldots \\[2mm] \dfrac{\partial Q'}{\partial w_{in}} = 0 \\[2mm] \dfrac{\partial Q'}{\partial \lambda} = 0 \end{cases},$$

so

$$w_{ij}^{*} = \frac{\displaystyle\sum_{k=1}^{m} w_{ij}^{k}}{m}, \ i = 1, \ldots, t.$$

Because $P_{7.1}$ is convex programming, w_{ij}^* is the optimal solution. (Convex programming is a nonlinear technique used in operational research.)

According to Theorem 7.1, the deviation of alternative weight and group preference based on the judgement matrix of decision maker k in stage i can be expressed as:

$$d_i^k = \sum_{j=1}^{n} (w_{ij}^* - w_{ij}^k)^2, \, i = 1, \dots, t, k = 1, \dots, m \qquad (7.3)$$

For decision maker k, the average value of stage is $d^k = \sqrt[t]{\prod_{i=1}^{t} d_i^k}, k = 1, \dots, m$, if some decision makers' preferences approach group preference and the deviation of group preference and individual preference is the smallest, one can let

$$e_{d^k} = \frac{\dfrac{\min\limits_{k=1,\dots,m} d^k}{d^k}}{\sum\limits_{k=1}^{m} \dfrac{\min\limits_{k=1,\dots,m} d^k}{d^k}}$$

as the weight based on the deviation between group and individual preference.

Because the value of e_{CR}^k and e_{d^k} has no consistent meaning, one uses Formula (7.4) to calculate the final weight of decision makers. Suppose that $\beta_d + \beta_{CR} = 1$, β_d is the offset distance and β_{CR} is the adjustment factor of the consistency ratio. If $\beta_d \geq \beta_{CR}$, the concentration degree of preference is more important in confirming the weight of decision makers; otherwise, the consistency degree when the decision makers make judgment is more important. Generally, suppose that $\beta_d = \beta_{CR}$ in the actual decision-making process.

$$pe^k = \frac{\sqrt{(\beta_d e_d^k)(\beta_{CR} e_{CR}^k)}}{\sum\limits_{k=1}^{m} \sqrt{(\beta_d e_d^k)(\beta_{CR} e_{CR}^k)}}, \, k = 1, \dots, m \qquad (7.4)$$

Definition 7.2

Suppose that pe^k is the comprehensive weight of decision makers. Modify the group preference w_{ij} of stage $i \in \{1, \dots, t\}$ after one obtains pe^k as $\mathrm{P}_{7.2}$.

$$\min \sum_{j,k} pe^k (w_{ij} - w_{ij}^{\,k})^2, \ i = 1, \dots, t$$

$$\begin{cases} \displaystyle\sum_{j=1}^{n} w_{ij} = 1, \ i = 1, \dots, t \\[2mm] w_{ij} \geq 0, \ i = 1, \dots, t, j = 1, \dots, n \end{cases}$$

Theorem 7.2

The optimization result of group preference based on $\mathrm{P}_{7.2}$ is

$$w_{ij}' = \sum_{k=1}^{m} pe^k w_{ij}^{\,k}, \ i = 1, \dots, t, \ j = 1, \dots, n.$$

Proof

The same as Theorem 7.1.

7.2.3 Weight of the Stage Based on Prior Information

Based on the weight of alternative j in stage i obtained by the former section, $w_{ij}, i = 1, \dots, t, \ j = 1, \dots, n$, this section researches how to aggregate the preferences of different t stages. First, one supposes that the weight of each stage is equal, so average group preference based on stage t can be expressed as Formula (7.5) and one can obtain $\sum_{j=1}^{n} w_j = 1$ easily.

$$w_j = \dfrac{\displaystyle\sum_{i=1}^{t} w_{ij}}{t}, \ j = 1, \dots, n \qquad (7.5)$$

Generally, the following principles should be considered in setting the weight p_i, $i = 1, \ldots, t$ of the decision-making stage:

1. The closer the decision-making stage, the more it will reflect the development of alternatives and decision makers' new acknowledgment. The conclusion is $p_i \leq p_{i+1}$.
2. The weight ratio factor α_i among the stages can also be set in the multiple-stage decision-making process, $\alpha_i p_i \leq p_{i+1}$, $\alpha_i \geq 1$.
3. The goal of decision makers' weight is to make the deviation of each stage preference and total preference be the smallest $[d_i = \sum_j p_i(w_j - w_{ij})^2, i = 1, \ldots, t]$.

One can establish the model of weight p_i as $P_{7.3}$, w_j is known—obtained by Formula (7.5).

$$\min \sum_{i,j} p_i(w_j - w_{ij})^2$$

$$\begin{cases} \alpha_i p_i \leq p_{i+1}, \alpha_i \geq 1 \\ \sum_{i=1}^{t} p_i = 1 \\ p_i \geq 0, i = 1, \ldots, t \end{cases}$$

Then establish model $P_{7.4}$ to confirm the final order w'_j, p_i is obtained according to $P_{7.3}$.

$$\min \sum_{i,j} p_i(w'_j - w_{ij})^2$$

$$P_{7.4} \begin{cases} \sum_{j=1}^{n} w'_j = 1 \\ w'_j \geq 0, j = 1, \ldots, n \end{cases}$$

Theorem 7.3

The optimized solution of $P_{7.4}$ can be expressed as

$$w_j'^* = \sum_{i=1}^{t} p_i w_{ij} \,.$$

Proof

The same as Theorem 7.1.

7.2.4 Example Analysis

A department must choose a leader and there are four candidates. Their comprehensive abilities must be examined by an examination team that consists of five experts conducting three assessments about the candidates, and the data are as follows. Which is the best?

The first-stage data:

$$\text{MD1:} \begin{bmatrix} 0.5 & 0.2 & 0.75 & 0.67 \\ 0.8 & 0.5 & 0.8 & 0.67 \\ 0.25 & 0.2 & 0.5 & 0.33 \\ 0.33 & 0.33 & 0.67 & 0.5 \end{bmatrix},$$

$$\text{MD2:} \begin{bmatrix} 1 & 1 & 3 & 1 \\ 1 & 1 & 3 & 2 \\ 1/3 & 1/3 & 1 & 1/2 \\ 1 & 1/2 & 2 & 1 \end{bmatrix},$$

$$\text{MD3:} \begin{bmatrix} 0.5 & 0.25 & 0.5 & 0.33 \\ 0.75 & 0.5 & 0.67 & 0.67 \\ 0.5 & 0.33 & 0.5 & 0.5 \\ 0.67 & 0.33 & 0.5 & 0.5 \end{bmatrix},$$

$$\text{MD4: } \begin{bmatrix} 1 & 1/3 & 3 & 2 \\ 3 & 1 & 5 & 2 \\ 1/3 & 1/5 & 1 & 2 \\ 1/2 & 1/2 & 1/2 & 1 \end{bmatrix},$$

$$\text{MD5: } \begin{bmatrix} 1 & 2 & 1 & 3 \\ 1/2 & 1 & 1/2 & 1 \\ 1 & 2 & 1 & 2 \\ 1/3 & 1 & 1/2 & 1 \end{bmatrix}.$$

The second-stage data:

$$\text{MD1: } \begin{bmatrix} 1 & 1/3 & 1 & 1/2 \\ 3 & 1 & 2 & 1 \\ 1 & 1/2 & 1 & 1/3 \\ 2 & 1 & 3 & 1 \end{bmatrix},$$

$$\text{MD2: } \begin{bmatrix} 0.5 & 0.33 & 0.67 & 0.25 \\ 0.67 & 0.5 & 0.8 & 0.67 \\ 0.33 & 0.2 & 0.5 & 0.25 \\ 0.75 & 0.33 & 0.75 & 0.5 \end{bmatrix},$$

$$\text{MD3: } \begin{bmatrix} 1 & 5 & 3 & 2 \\ 1/5 & 1 & 1/2 & 1/2 \\ 1/3 & 2 & 1 & 2 \\ 1/2 & 2 & 1/2 & 1 \end{bmatrix},$$

$$\text{MD4: } \begin{bmatrix} 0.5 & 0.25 & 0.33 & 0.75 \\ 0.75 & 0.5 & 0.2 & 0.75 \\ 0.67 & 0.8 & 0.5 & 0.8 \\ 0.33 & 0.25 & 0.2 & 0.5 \end{bmatrix},$$

$$\text{MD5:} \begin{bmatrix} 1 & 1/4 & 2 & 3 \\ 4 & 1 & 2 & 5 \\ 1/2 & 1/2 & 1 & 3 \\ 1/3 & 1/5 & 1/3 & 1 \end{bmatrix}.$$

The third-stage data:

$$\text{MD1:} \begin{bmatrix} 1 & 1 & 1/3 & 1/2 \\ 1 & 1 & 1/2 & 1/3 \\ 3 & 2 & 1 & 2 \\ 2 & 3 & 1/2 & 1 \end{bmatrix},$$

$$\text{MD2:} \begin{bmatrix} 0.5 & 0.75 & 0.83 & 0.75 \\ 0.25 & 0.5 & 0.67 & 0.5 \\ 0.17 & 0.33 & 0.5 & 0.33 \\ 0.25 & 0.5 & 0.67 & 0.5 \end{bmatrix},$$

$$\text{MD3:} \begin{bmatrix} 1 & 3 & 6 & 7 \\ 1/3 & 1 & 2 & 2 \\ 1/6 & 1/2 & 1 & 1 \\ 1/7 & 1/2 & 1 & 1 \end{bmatrix},$$

$$\text{MD4:} \begin{bmatrix} 1 & 1/3 & 2 & 3 \\ 3 & 1 & 2 & 5 \\ 1/2 & 1/2 & 1 & 2 \\ 1/3 & 1/5 & 1/2 & 1 \end{bmatrix},$$

$$\text{MD5:} \begin{bmatrix} 0.5 & 0.25 & 0.33 & 0.5 \\ 0.75 & 0.5 & 0.5 & 0.67 \\ 0.67 & 0.5 & 0.5 & 0.67 \\ 0.5 & 0.33 & 0.33 & 0.5 \end{bmatrix}.$$

Step 1: Transform the multiple judgement preference matrix given by the decision makers in each step into a reciprocal judgment matrix and then calculate the consistency ratio CR and the weight. The result shows that the decision makers' judgment consistency is good in each stage, except the consistency of decision maker 4 in steps 1 and 2 is greater than 0.1. Because this section confirms the decision maker's weight based on the consistency of the decision maker, one does not deal with CR that is greater than 0.1.

Step 2: Confirm the weight of decision makers and aggregation of group preference.

1. Confirm the weight $e_{CR}{}^k$ of decision makers based on the consistency judgment level of decision makers. According to Formula (7.2) and Definition 7.1, one obtains the result that the judgment consistency of decision maker 3 is the best; the weight is 0.356. The ranking of the judgment consistency of the decision makers is 2, 5, 1, and 4. Decision makers 2 and 3 achieved better consistency than decision makers 1 and 4. Although the average consistency value of the last two decision makers is lower than 0.1, the consistency ratio is obviously higher than decision makers 2 and 3.
2. Confirm the weight of decision makers based on the difference of personal and group judgment. First, one supposes that the weight is equal and then calculates group preference in each stage according to $P_{7.1}$. Then calculate the difference between decision maker judgment preference and group preference. Calculate the weight of decision makers in each stage according to Formula (7.3). The result is the difference between decision maker 5, and group preference is the minimum and decision maker 3 is the maximum.
3. Confirm the comprehensive weight of decision makers. Calculate the weight based on Formula (7.4), as in Table 7.1.

    One can see that $e_{CR}{}^k$ is large, while e_{q^k} is not. Take MD3, for example. The highest weight based on consistency is 0.356 but the deviation between this decision maker's judgment preference and group preference is not the smallest. The method combines these two weights using Formula (7.4).

Table 7.1   Comprehensive Weight of Decision Makers

	MD1	MD2	MD3	MD4	MD5
e_{CR}^{k}	0.092	0.315	0.356	0.037	0.200
$e_{d^{*}}$	0.220	0.240	0.123	0.159	0.258
pe^{k}	0.153	0.296	0.225	0.083	0.244

Table 7.2   Sum of Squares of Each Stage Order Deviation

	ALTERNATIVE 1	ALTERNATIVE 2	ALTERNATIVE 3	ALTERNATIVE 4	d_i
Step 1	0.0262	0.0827	0.0369	0.0093	0.155
Step 2	0.0852	0.1054	0.1075	0.069	0.367
Step 3	0.203	0.089	0.084	0.029	0.405

4. The calculation of group preference in each stage based on the weight of decision makers ($P_{7.2}$). The weight of alternatives changes in different stages. The weight of alternative 2 becomes smaller, the weight of alternative 3 is improved, and alternative 1 fluctuates, but overall it has an increasing trend; the alternative has a decreasing trend.

Step 3: Confirm the weight of each stage and modify the weight of the alternative. One supposes that the weights are equal in the beginning. The difference between each stage weight and the comprehensive weight can be obtained by calculating the order of alternative based on multiple stages, as in Table 7.2. It is obvious that the deviation between the three decision-making stages and trial stage group preference is not the same; stage 1 is the smallest but it is increasing.

Based on $P_{7.3}$

$$\min z = 0.155\,p_{1} + 0.367\,p_{2} + 0.405\,p_{3}$$

$$\text{s.t.} \begin{cases} p_{3} \geq p_{2} \geq p_{1} \\ p1 + p2 + p3 = 1 \\ p1, p2, p3 \geq 0 \end{cases} \quad ,$$

one can obtain $p_{1} = p_{2} = p_{3} = 1/3$. Generally, the judgment that is closer to the ultimate decision stage should be emphasized, but the weight of three stages is balanced after modifying the distance of comprehensive deviation. According to the final result of three stages, $w_1 = 0.297$, $w_1 = 0.320$, $w_1 = 0.193$, $w_1 = 0.192$, the result demonstrates that the order is alternative 2, alternative 1, alternative 3, and alternative 4. The distance of the last two alternatives is the shortest.

Group decision making based on multiple stages is worth studying. This section puts forward the model of confirming the weight of decision makers and decision-making stage and then establishes the aggregated model of multidecision stages, multiple decision makers, and multiclass judgment preference. Last, this section proves all characteristics of this method. The method is easily understood. Future research will focus on how to aggregate uncertain preferences of multiple decision stages and multiclasses.

7.3 Aggregation Method of Uncertain Preference with Multiple Stages

This section researches the aggregation method of uncertain preference with multiple stages, including interval number reciprocal judgment matrix, interval number reciprocal judgment matrix, interval number reciprocal judgment matrix, and order of interval number preference.

7.3.1 Weight of the Uncertain Preference

The interval number form of the weight of uncertain preference is $[wa_i^L, wa_i^U], [wb_i^L, wb_i^U], [we_i^L, we_i^U], [wf_i^L, wf_i^U], i = 1, \ldots, n$. This formula demonstrates the weight of uncertain preference based on decision maker k.

7.3.2 Weight of the Decision Maker Based on Interval Number

Suppose that the comprehensive preference of the group is $[w_i^{gL}, w_i^{gU}]$ and define the function based on the deviations among the four weights is $z = \sum_{i,k} \beta_k (| w_i^{gL} - w_i^{kL} | + | w_i^{gU} - w_i^{kU} |)$, β_k, $k = 1, \ldots, m$, m is the weight of each decision maker, and w_i^{kL} and w_i^{kU} are the top and bottom limitation of decision maker k based on one of the four preference forms. The smaller z indicates the smaller distance between the group preference and individual preference. One can set $P_{7.5}$ in order to obtain the smaller z.

$$\text{P}_{7.5} \quad \min z = \sum_{i,k} \beta_k \left(\left| w_i^{gL} - w_i^{kL} \right| + \left| w_i^{gU} - w_i^{kU} \right| \right)$$

$$\text{s.t.} \quad 0 \le w_i^{gL} \le w_i^{gU} \le 1$$

One transforms $\text{P}_{7.5}$ to $\text{P}_{7.6}$ in order to solve conveniently; pLl_{ik}, pLu_{ik}, pUl_{ik}, pUu_{ik} are deviation variables.

$$\min z = \sum_{i,k} \beta_k (pLl_{ik} + pLu_{ik} + pUl_{ik} + pUu_{ik})$$

$$\text{P}_{7.6} \quad \text{s.t.} \begin{cases} w_i^{gL} - w_i^{kL} + pLl_{ik} - pLu_{ik} = 0, \ i = 1, \dots , n, \ k = 1, \dots , m \\[2mm] w_i^{gU} - w_i^{kU} + pUl_{ik} - pUu_{ik} = 0, \ i = 1, \dots , n, \ k = 1, \dots , m \\[2mm] 0 \le w_i^{gL} \le w_i^{gU} \le 1, \ i = 1, \dots , n \\[2mm] pLl_{ik}, \ pLu_{ik}, \ pUl_{ik}, \ pUu_{ik} \ge 0, \ i = 1, \dots , n, \ k = 1, \dots , m \end{cases}$$

β_k is the uncertain variable, so the model is a nonlinear programming. In this section one supposes that

$$\beta_k = \frac{1}{m}, \ k = 1, \dots , m$$

and the weight of each decision maker is equal. One obtains w_i^{gL}, w_i^{gU} according to $\text{P}_{7.6}$. Then modify the weight of the decision maker according to the deviation between individual preferences w_i^{kL}, w_i^{kU} and w_i^{gL}, w_i^{gU}.
If

$$\beta_k = \frac{1}{m}, \ k = 1, \dots , m$$

in $\text{P}_{7.6}$, then order w_i^{kL} (or w_i^{kU}), the optimal solution of w_i^{gL} is $w_i^{gL*} = w_i^{kL}$, $k = (m + 1)/2$ if m is odd; otherwise, $w_i^{gL*} \in (w_i^{m/2L}, w_i^{m/2+1L})$. The solution of w_i^{gU*} is $w_i^{gU*} = w_i^{kU}$, $k = (m + 1)/2$ and $w_i^{gU*} \in (w_i^{m/2U}, w_i^{m/2+1U})$.

Definition 7.3

Suppose that the weight of decision maker k to alternative i is $[w_i^{kL}, w_i^{kU}]$, $i = 1, \ldots, n$, $k = 1, \ldots, m$, the group comprehensive preference of alternative i is $[w_i^{gL}, w_i^{gU}]$ and

$$\xi_{gk} = \sum_{i=1}^{n} \frac{\min_{k} \min_{i}\{L_{gk}(i)\} + \rho \max_{k} \max_{i}\{L_{gk}(i)\}}{L_{gk}(i) + \rho \max_{k} \max_{i}\{L_{gk}(i)\}} \Bigg/ n$$

is the decision-making correlation coefficient of decision maker k and group preference.

$$L_{gk}(i) = \frac{\sqrt{(w_i^{kL} - w_i^{gL})^2 + (w_i^{kU} - w_i^{gU})^2}}{\sqrt{2}}$$

is the distance of interval number $[w_i^{kL}, w_i^{kU}]$ and $[w_i^{gL}, w_i^{gU}]$, $\rho \in (0, 1)$ is a discrimination coefficient.

Definition 7.3 is derived from the concept of interval number correlation. This index can express the satisfaction of group preference based on the deviation between the preference weight of the decision maker and the comprehensive preference of the decision-making group. The deviations of different ranks should be considered, so this section puts forward using Spearman's rank correlation coefficient to measure the deviation.

Spearman's rank correlation coefficient can be used to measure the deviation of $rank(w_i)$ and $rank(w_i')$. Because the preference weights of decision maker and group are both interval numbers, generally the method cannot order them from small to large. So, order $[w_i^{kL}, w_i^{kU}]$, $i = 1, \ldots, n$, $k = 1, \ldots, m$ and $[w_i^{gL}, w_i^{gU}]$, $i = 1, \ldots, n$ separately, then compare the interval number two by two and make a preference matrix. Solve the matrix and obtain $w_i^{k'}$, $w_i^{g'}$, $i = 1, \ldots, n$. Last, confirm the order of interval number according to $w_i^{k'}$, $w_i^{g'}$, $i = 1, \ldots, n$.

Definition 7.4

Suppose that $w_i^{k'}$ is the weight based on decision maker k and $w_i^{g'}$, $i = 1, \ldots, n$ is the weight based on comprehensive preference,

$$SRC^{kg} = 1 - \frac{6\sum [rank(w_i^{k'}) - rank(w_i^{g'})]^2}{n(n^2 - 1)}$$

$-1 \leq SRC^{kg} \leq 1$ is the order correlation coefficient between decision maker k and comprehensive preference. If the order is perfect correlation, $SRC^{kg} = 1$; otherwise, $SRC^{kg} = -1$.

Calculate SRC^{kg} and ξ_{gk} separately. One uses the geometric mean to express the index value of decision maker k.

$$\beta_k = \frac{\sqrt{\xi_{gk} SRC^{kg}}}{\sum \sqrt{\xi_{gk} SRC^{kg}}}$$

is the weight of decision maker k in this section. The two index values in β_k have a certain correlation but not an inevitable correlation. Then calculate the group preference w_i^{gL}, w_i^{gU} based on β_k and $P_{7.6}$.

7.3.3 Weight Model Based on Least Deviation

Suppose that $[w_{tj}^{gL}, w_{tj}^{gU}]$, $t = 1, \ldots, T, j = 1, \ldots, n$ is the weight of alternative j in stage t. This section researches how to aggregate the preference of T stages. First, suppose that the weights are equal, so the average preference $[w_i^L, w_i^U]$ based on T stages can be expressed as Formula (7.6).

$$w_j^L = \frac{\sum_{t=1}^{T} w_{tj}^{gL}}{T}, \quad w_j^U = \frac{\sum_{t=1}^{T} w_{tj}^{gU}}{T}, \quad j = 1, \ldots, n \quad (7.6)$$

Generally, the following methods and principles should be considered in setting p_t, $t = 1, \ldots, T$:

1. The closer the decision-making stage, the more it will reflect the development of alternatives and decision makers' acknowledgment. The conclusion is $p_t \leq p_{t+1}$.

2. The weight ratio factor α_t among the stages can also be set in the multiple-stage decision-making process, $\alpha_t p_t \leq p_{t+1}$, $\alpha_t \geq 1$.
3. The target of decision makers' weight is makes the deviation of each stage preference and total preference be the smallest.

One can establish the model of weight p_t as $P_{7.7}$,

$$w_j^L = \frac{\sum_{t=1}^{T} w_{tj}^{gL}}{T}, \quad w_j^U = \frac{\sum_{t=1}^{T} w_{tj}^{gU}}{T}, j = 1, \dots, n$$

The values of w_j^L and w_j^U are known.

$$\min \sum_{t,j} p_t \{ \left| (w_j^L - w_{tj}^{gL}) \right| + \left| (w_j^U - w_{tj}^{gU}) \right| \}$$

$$\begin{cases} \alpha_t p_t \leq p_{t+1}, \alpha_t \geq 1 \\ \sum_{t=1}^{T} p_t = 1 \\ p_t \geq 0, t = 1, \dots, T \end{cases}$$

Then establish model $P_{7.8}$ to confirm the final order. p_t is obtained according to $P_{7.8}$.

$$P_{7.8} : \quad \min \sum_{t,j} p_t \{ \left| (w_j^{L'} - w_{tj}^{gL}) \right| + \left| (w_j^{U'} - w_{tj}^{gU}) \right| \}$$

$$w_j^{L'}, w_j^{U'} \geq 0, j = 1, \dots, n$$

The method can be summarized as the following steps:

Step 1: Calculate the alternative order of uncertain preference of the decision-making group.
Step 2: Set the weight of the decision-making group and calculate the comprehensive preference of the group in some decision-making stages.
Step 3: Confirm the weight of the decision-making stages and calculate the comprehensive preference of the group of multiple stages.

7.3.4 Advertisement Effect Evaluation of Multiple-Stage Preference Aggregation

7.3.4.1 Background   In China, domestic automobile production and marketing in 2006 became the fastest growing after 2003. The automobile market has five characteristics: rapid growth of production and marketing, expansion of automobile market, growth in production of commercial vehicles, consistent profits, and development of related enterprises. The growth of the automotive industry also makes competition more fierce. H company is a key automobile enterprise; its annual output is half a million vehicles and 600,000 engines and it has produced and sold more than 2 million vehicles. The advertisement policy of H is as follows: in the print media using newspapers and magazines to propagandize, with the key point being newspapers; outdoor media advertising focuses on developed economic regions, such as Beijing, Shanghai, and so on; H company makes significant investment in TV media, which also involves radio media and subway media, and the important channels are A2, A5, A10, and A12. The key point of this section is evaluating the four channels and their deviations.

7.3.4.2 Criteria and Basic Information   The following data should be provided in the evaluation process. First the rate of coverage: whether the audience can contact the advertisement or not; one calls this advertising coverage rate usually. The index of whether an audience is exposed to an advertisement (advertising coverage rate) is measured. The index is a simple measure based on the premise that consumers may or may not watch advertisements on television. Second is the sum of impression percentage. Third is exposure frequency: the average exposure times to potential consumers in a certain period. The last is the advertising cost per thousand, which means the average cost to convey the advertisement to 1,000 people.

H company established the following rule for evaluating advertisements. They evaluate all advertisements every season and perform a comprehensive evaluation every year. The evaluation group includes one expert from the advertising company, the sales manager and vice manager, a planning department manager, and a planning manager from the advertising company. The evaluation group evaluates every season according to the data that are provided by H, but the decision makers have different opinions about the data, so some qualitative knowledge and experience are involved in the evaluation process. Thus the advertisement effect may be compared by the pairwise model: The four channels can be expressed as A2, A5, A10, and A12.

The first stage:

A: [0.2, 0.3]   [0.25, 0.3]   [0.4, 0.6]   [0.1, 0.2];

B: $\left[[3, 4] \quad [2, 3] \quad [1, 2] \quad [3, 4] \right]$;

C: $\left[[0.1, 0.3] \quad [0.2, 0.4] \quad [0.5, 0.6] \quad [0.1, 0.2] \right]$;

D:
$$\begin{bmatrix} [1, 1] & [1/2, 2/3] & [1/2, 1] & [2, 3] \\ & [1, 1] & [1/2, 1] & [1/5, 1/3] \\ & & [1, 1] & [2, 4] \\ & & & [1, 1] \end{bmatrix};$$

E:
$$\begin{bmatrix} [0.5, 0.5] & [0.2, 0.4] & [0.2, 0.3] & [0.5, 0.6] \\ & [0.5, 0.5] & [0.4, 0.5] & [0.2, 0.3] \\ & & [0.5, 0.5] & [0.6, 0.8] \\ & & & [0.5, 0.5] \end{bmatrix}.$$

The second stage:

A: $\left[[2, 3] \quad [1, 3] \quad [3, 4] \quad [4, 4] \right]$;

B: $\left[[0.1, 0.2] \quad [0.4, 0.5] \quad [0.2, 0.3] \quad [0.1, 0.2] \right]$;

C:
$$\begin{bmatrix} [1, 1] & [1/3, 1/2] & [1/3, 1] & [1, 2] \\ & [1, 1] & [2, 3] & [4, 6] \\ & & [1, 1] & [2, 3] \\ & & & [1, 1] \end{bmatrix};$$

D:
$$\begin{bmatrix} [1, 1] & [1/4, 1/2] & [1/3, 1/2] & [2, 3] \\ & [1, 1] & [3, 4] & [5, 6] \\ & & [1, 1] & [1, 2] \\ & & & [1, 1] \end{bmatrix};$$

E: $\left[[0.2, 0.3] \quad [0.5, 0.6] \quad [0.2, 0.3] \quad [0.1, 0.2] \right]$.

The third stage:

$$
A: \begin{bmatrix}
[1, 1] & [1/4, 1/2] & [1/2, 1] & [1, 3] \\
 & [1, 1] & [2, 4] & [5, 6] \\
 & & [1, 1] & [1, 2] \\
 & & & [1, 1]
\end{bmatrix};
$$

B: $\begin{bmatrix} [0.15, 0.2] & [0.5, 0.6] & [0.3, 0.4] & [0.1, 0.2] \end{bmatrix}$;

C: $\begin{bmatrix} [0.1, 0.2] & [0.4, 0.5] & [0.2, 0.3] & [0.1, 0.2] \end{bmatrix}$;

$$
D: \begin{bmatrix}
[0.5, 0.5] & [0.3, 0.4] & [0.4, 0.5] & [0.6, 0.7] \\
 & [0.5, 0.5] & [0.7, 0.8] & [0.6, 0.7] \\
 & & [0.5, 0.5] & [0.5, 0.6] \\
 & & & [0.5, 0.5]
\end{bmatrix};
$$

E: $\begin{bmatrix} [2, 3] & [1, 2] & [3, 4] & [3, 4] \end{bmatrix}$.

The fourth stage:

A: $\begin{bmatrix} [0.2, 0.3] & [0.4, 0.5] & [0.2, 0.3] & [0.1, 0.2] \end{bmatrix}$;

B: $\begin{bmatrix} [3, 4] & [1, 2] & [2, 3] & [3, 4] \end{bmatrix}$;

$$
C: \begin{bmatrix}
[1, 1] & [1/3, 1/2] & [1/2, 1] & [1, 2] \\
 & [1, 1] & [4, 5] & [5, 6] \\
 & & [1, 1] & [1, 2] \\
 & & & [1, 1]
\end{bmatrix};
$$

$$
D: \begin{bmatrix}
[0.5, 0.5] & [0.3, 0.5] & [0.2, 0.3] & [0.5, 0.6] \\
 & [0.5, 0.5] & [0.6, 0.8] & [0.6, 0.9] \\
 & & [0.5, 0.5] & [0.5, 0.6] \\
 & & & [0.5, 0.5]
\end{bmatrix};
$$

E: $\begin{bmatrix} [0.1, 0.2] & [0.4, 0.5] & [0.1, 0.2] & [0.2, 0.3] \end{bmatrix}$.

7.3.4.3 Result and Analysis

Step 1: Calculate the weight of alternative given by decision makers at each stage ($\gamma = 0.05$), as in Tables 7.3–7.6.

Step 2: Confirm the weight of decision makers based on decision makers' correlation and preference order. Supposing that

Table 7.3 Alternative Order of Preference Information in the First Stage

THE FIRST STAGE	ALTERNATIVE 1	ALTERNATIVE 2	ALTERNATIVE 3	ALTERNATIVE 4
Expert A	[0.2, 0.3]	[0.25, 0.3]	[0.4, 0.6]	[0.1, 0.2]
Expert B	[0.0096, 0.1699]	[0.1699, 0.3301]	[0.3301, 0.4904]	[0.0096, 0.1699]
Expert C	[0.1, 0.3]	[0.2, 0.4]	[0.5, 0.6]	[0.1, 0.2]
Expert D	[0.1976, 0.2417]	[0.2113, 0.3063]	[0.3874, 0.4507]	[0.098, 0.1209]
Expert E	[0.1538, 0.1792]	[0.2336, 0.2729]	[0.3593, 0.4145]	[0.1620, 0.2443]

Table 7.4 Alternative Order of Preference Information in the Second Stage

THE SECOND STAGE	ALTERNATIVE 1	ALTERNATIVE 2	ALTERNATIVE 3	ALTERNATIVE 4
Expert A	[0.1698, 0.33010]	[0.1698, 0.4903]	[0.0096, 0.1698]	[0.0096, 0.0096]
Expert B	[0.1, 0.2]	[0.4, 0.5]	[0.2, 0.3]	[0.1, 0.2]
Expert C	[0.16, 0.2222]	[0.4444, 0.5455]	[0.1818, 0.25]	[0.0833, 0.12]
Expert D	[0.1426, 0.1860]	[0.5411, 0.5717]	[0.1754, 0.1906]	[0.0904, 0.0954]
Expert E	[0.2, 0.3]	[0.5, 0.6]	[0.2, 0.3]	[0.1, 0.2]

Table 7.5 Alternative Order of Preference Information in the Third Stage

THE THIRD STAGE	ALTERNATIVE 1	ALTERNATIVE 2	ALTERNATIVE 3	ALTERNATIVE 4
Expert A	[0.1351, 0.2000]	[0.5000, 0.6000]	[0.1471, 0.2162]	[0.0909, 0.1176]
Expert B	[0.15, 0.2]	[0.5, 0.6]	[0.3, 0.4]	[0.1, 0.2]
Expert C	[0.1, 0.2]	[0.4, 0.5]	[0.2, 0.3]	[0.1, 0.2]
Expert D	[0.1862, 0.2119]	[0.4244, 0.4435]	[0.1819, 0.1927]	[0.1764, 0.1889]
Expert E	[0.1699, 0.3301]	[0.3301, 0.4904]	[0.0096, 0.1699]	[0.0096, 0.1699]

Table 7.6 Alternative Order of Preference Information in the Fourth Stage

THE FOURTH STAGE	ALTERNATIVE 1	ALTERNATIVE 2	ALTERNATIVE 3	ALTERNATIVE 4
Expert A	[0.2, 0.3]	[0.4, 0.5]	[0.2, 0.3]	[0.1, 0.2]
Expert B	[0.0096, 0.1698]	[0.3301, 0.4904]	[0.1698, 0.3301]	[0.0096, 0.1699]
Expert C	[0.1470, 0.1916]	[0.5594, 0.5884]	[0.1382, 0.1471]	[0.0951, 0.1177]
Expert D	[0.1655, 0.1754]	[0.3897, 0.4041]	[0.2533, 0.2655]	[0.1682, 0.1788]
Expert E	[0.1, 0.2]	[0.4, 0.5]	[0.1, 0.2]	[0.2, 0.3]

Table 7.7 Confirm the Weight of Decision Makers Based on Correlation

STAGE	FIRST	SECOND	THIRD	FOURTH	GEOMETRIC AVERAGE CORRELATION	WEIGHT
Decision maker 1	0.6152	0.4686	0.6768	0.7267	0.6136	0.1985
Decision maker 2	0.4926	0.6736	0.6767	0.5114	0.5821	0.1883
Decision maker 3	0.5749	0.8862	0.7429	0.5894	0.6872	0.2223
Decision maker 4	0.5878	0.6756	0.7030	0.5918	0.6375	0.2062
Decision maker 5	0.5665	0.6240	0.4776	0.6278	0.5706	0.1846

Table 7.8 Confirm the Weight of Decision Makers Based on Preference Order

DECISION-MAKING STAGE	FIRST	SECOND	THIRD	FOURTH	GEOMETRIC AVERAGE ORDER	WEIGHT
Decision maker 1	1.0000	0.8000	1.0000	0.9500	0.9337	0.2137
Decision maker 2	0.9500	0.9500	1.0000	0.9500	0.9623	0.2203
Decision maker 3	1.0000	1.0000	0.9500	0.8000	0.9337	0.2137
Decision maker 4	1.0000	1.0000	0.8000	0.8000	0.8944	0.2047
Decision maker 5	0.8000	0.9500	0.6500	0.3500	0.6448	0.1476

the weight is the same in the beginning, confirm the weight in each stage, then the estimated comprehensive preference of each stage can be obtained. Calculate the geometric average correlation and geometric average order of each stage and then obtain the relative weight, as shown in Tables 7.7 and 7.8.

The comprehensive weight of decision makers: decision maker 1: 0.2064, decision maker 2: 0.2040, decision maker 3: 0.2184, decision maker 4: 0.2059, decision maker 5: 0.1654. The result demonstrates that the weights of decision makers are almost at the same level. It also demonstrates that the judgment level of decision makers and the view of the group are consistent, although the deviation between decision maker 5 and the group is greater and its weight is lower. The comprehensive preference of decision makers is listed in Table 7.9.

Step 3: Confirm the weight of the decision-making stage. First suppose that the weight is equal. According to $|(w_j^L - w_{ij}^{gL}) + (w_j^U - w_{ij}^{gU})|$, one can obtain the deviation of each stage: 0.737025, 0.396225, 0.285225, and 0.224925. Suppose that the weights of stages 1, 2, 3, and 4 satisfy the following requirement: $p_1 \geq 0.05, p_2 \geq 2 p_1, p_3 \geq 2 p_2, p_4 \geq 2 p_3$, then one can obtain the weight of four stages $p_1 \geq 0.05, p_2 \geq 2 p_1$,

Table 7.9 Comprehensive Group Preference of Each Stage Based on the Weight of Decision Makers

	ALTERNATIVE 1	ALTERNATIVE 2	ALTERNATIVE 3	ALTERNATIVE 4
Step 1	[0.1538, 0.2417]	[0.2113, 0.3063]	[0.3874, 0.4904]	[0.100, 0.2]
Step 2	[0.16, 0.2222]	[0.4444, 0.5455]	[0.1818, 0.25]	[0.0096, 0.12]
Step 3	[0.15, 0.2]	[0.4244, 0.5]	[0.1819, 0.2162]	[0.1, 0.1889]
Step 4	[0.147, 0.1916]	[0.4, 0.5]	[0.1698, 0.2655]	[0.1, 0.1788]

$p_3 \geq 2\,p_2, p_4 \geq 2\,p_3$. The weight of the fourth is greatest because the sum of deviation of this stage is the smallest and the decision of this stage is newer. Based on the weight of stages, one can obtain the comprehensive group preference, [0.147, 0.1916], [0.4, 0.5], [0.1698, 0.2655], [0.1, 0.1788]. Then decision makers can order the weights of interval number form and confirm the final order of the alternative. Advertising plays an important role in the competitive automotive industry and evaluating advertising effect is an important marketing activity. It is essentially a typical multiple-stage and -index problem. This chapter puts forward a group evaluation framework and model based on uncertain preference. This method is easy to understand and has practical value.

7.4 Summary and Future Research

This chapter researches the aggregation method based on multiple-stages preference and put forward the model of confirming the weight. How to confirm the weight of stages, decision makers, and decision-making indices is important in the multiple-stage decision-making process. Therefore, the following problems should be considered.

1. Research on the quality of multiple-stages information. Because multistage decision making includes many stages, the time and the quality of information may be not consistent. Based on this, the key point of multistage preference aggregation is how to evaluate the quality of information in each stage.
2. The multistage decision-making problem adds time latitude to traditional decision making and this makes the problem more complex. So this problem should be paid attention to in the process of multiple-stage preference aggregation.

8

An Extension of TOPSIS with Multiple-Stages Fuzzy Linguistic Evaluation for Group Decision Making

8.1 Introduction

The decision maker (DM) has to make decisions in a complex environment. He or she has to deal with the problem of uncertain information and has to respond to it rapidly. As one of the most important aspects of decision-making, multi-attribute decision making (MADA) is widely used. For example, the criteria of price, space ratio, floor space, location, and surroundings are considered in the purchase a house. One must consider equipment, production price, capability, management level, business credit standing, and enterprise culture to select the best partner. There exist many effective methods, such as the analytic hierarchy process, TOPSIS, ELECTRE [102, 103], PROMETHEE [104, 105], and so on.

The complex decision-making condition increases the decision-making difficulty. Many new theories and methods are put forward, such as stochastic decision making, fuzzy decision making, interval number method, and gray system theory [106, 107]. In addition, a vague set is put forward to solve the uncertainty. Due to the complexity of the alternative and the environment, the linguistic label is usually adopted, which concerns the consistency of linguistic judgment, the transformation of different linguistic scales, and the decision-making method in a group and in an incomplete information environment [108–111]. Moreover, the Dempster–Shafer theory [113–115] and the prospect theory [112] are also developed. We can conclude that the aggregation of multiple methods will solve the uncertainty problem.

In some cases, the evaluation information of the alternative in multiple stages should be aggregated. For example, one should consider the integration performance of every year in leader evaluation problem, though the evaluation process is proceeding in each year. In fact, it is a three-dimensional decision-making problem about time, criteria, and alternatives. Several methods of positive numbers are summarized in the literature [116–118], including the methods of two times weight, prompting and punishment, increasing disparity of vertical and transverse, and so on. However, the weight of criteria is supposed in advance. Due to its practicality and reliability, TOPSIS has been widely studied. TOPSIS in uncertainty is developed, such as application in fuzzy [119, 122, 125, 127], interval [121, 124], and interval-valued intuitionistic fuzzy numbers [120]. In addition, group TOPSIS has been suggested [126]. Roghani et al. [123] compared the different aggregation order and gave some suggestions.

According to the literature review, we can see that there is no relevant research on multiple-stage linguistic evaluation of TOPSIS. Two important points should be focused on. First, the measurement of the distance of linguistic variables between the ideal point and the alternatives should be studied. Although several literatures works have referred to the measurement of the distance of linguistic variable, the measurement of the distance is complex compared with the numerical value due to the fact that the relationship among linguistic scales is nonlinear. Second, it is difficult to obtain the weight of the stage that determines the rank of the alternative directly. In many cases, the change and development of stage has not been given much importance. As a result, the selection of an ideal point and measurement of the distance are researched. Moreover, one model of weight of stage is also suggested to solve the decision-making problem without weight information.

8.2 Main Methods and Results

8.2.1 Review of TOPSIS and Some Notes for Fuzzy Linguistic Evaluation

Let a_{ij} be the linguistic evaluation for criteria of alternative i, $i = 1, ..., n$. Suppose the weight of criteria j is w_j. First, review the TOPSIS for fuzzy numbers. Then, some properties of the method are discussed.

1. Linguistic variables and fuzzy numbers: Several linguistic sets with different granularities are suggested and studied in the literature. Some notes about the linguistic sets can be summarized. First, we can conclude that different linguistic sets should be used in different decision-making problems. One will obtain different results by using different linguistic sets with different granularities in one problem. Second, when the decision maker measures the alternative using the linguistic scale, he or she should keep up with the consistency and stabilization of the measurement. Most important, we should realize that in many cases the mathematical relation between the linguistic set and its fuzzy numbers may be set up according to the decision maker's subjective preference. For convenience, the linguistic variables a_{ij} for ratings and for importance weight w_j from Chou et al. [128] are adopted in this chapter, which are shown in Tables 8.1 and 8.2, respectively. The tables indicate the relationship between the linguistic variable and fuzzy number. For example, the "very poor" of the linguistic set measured by a trapezoidal fuzzy number (0, 0, 0, 20) shows the worst grade, which can be denoted as $\tilde{A} = [a, b, c, d]$. This style can also be used for quantitative and qualitative evaluation.

To defuzzify a fuzzy number, the weighting distance of trapezoidal fuzzy number $\tilde{A}$ is defined as

$$d(\tilde{A}) = \frac{a + b + c + d}{4} \tag{8.1}$$

The linguistic variables of Table 8.2 are used to express the importance of the weight for criteria. Based on Chou et al. [128], the weight for criteria j can be obtained as

Table 8.1   Linguistic Variables and Fuzzy Numbers for the Ratings

LINGUISTIC VARIABLES	FUZZY NUMBERS
Very poor (S1)	(0, 0, 0, 20)
Between very poor and poor (S2)	(0, 0, 20, 40)
Poor (S3)	(0, 20, 20, 40)
Between poor and fair (S4)	(0, 20, 50, 70)
Fair (S5)	(30, 50, 50, 70)
Between fair and good (S6)	(30, 50, 80, 100)
Good (S7)	(60, 80, 80, 100)
Between good and very good (S8)	(60, 80, 100, 100)
Very good (S9)	(80, 100, 100, 100)

Table 8.2   Linguistic Variables and Fuzzy Numbers for the Importance Weight

LINGUISTIC VARIABLES	FUZZY NUMBERS
Very low (VL)	(0, 0, 0, 3)
Low (L)	(0, 3, 3, 5)
Medium (M)	(2, 5, 5, 8)
High (H)	(5, 7, 7, 10)
Very high (VH)	(7, 10, 10, 10)

$$w_j = \frac{d(\varpi_j)}{\sum\limits_{j=1}^{n} d(\varpi_j)}, j = 1, \ldots, n \tag{8.2}$$

If $m = (m_1, m_2, m_3, m_4)$ and $n = (n_1, n_2, n_3, n_4)$ are two trapezoidal fuzzy numbers, the distance between them can be calculated using Formula (8.3) [127].

$$d(m, n) = \sqrt{\frac{(m_1 - n_1)^2 + (m_2 - n_2)^2 + (m_3 - n_3)^2 + (m_4 - n_4)^2}{4}} \tag{8.3}$$

2. Selection of positive ideal point (PIP) and negative ideal point (NIP): In general, the positive ideal point and negative ideal point are set up as Formula (8.4), where a_{ij} is the rating for alternative i based on criteria j.

$$\begin{cases} a_j^+ = \max_i(a_{ij}) \\[2mm] a_j^- = \min_i(a_{ij}) \end{cases} \tag{8.4}$$

Suppose that the PIP and NIP with trapezoidal fuzzy numbers are denoted as

$$\begin{cases} a_j^+ = (a_{j1}^+, a_{j2}^+, a_{j3}^+, a_{j4}^+) \\[2mm] a_j^- = (a_{j1}^-, a_{j2}^-, a_{j3}^-, a_{j4}^-) \end{cases} \tag{8.5}$$

So, the distance between the alternative i and PIP and NIP can be expressed as

$$\begin{cases} d(a_i, a^+) \\[4mm] = \sum_{j=1}^{n} w_j \sqrt{\dfrac{[(a_{ij1} - a_{j1}^+)^2 + (a_{ij2} - a_{j2}^+)^2 + (a_{ij3} - a_{j3}^+)^2 + (a_{ij4} - a_{j4}^+)^2]}{4}} \\[6mm] d(a_i, a^-) \\[4mm] = \sum_{j=1}^{n} w_j \sqrt{\dfrac{(a_{ij1} - a_{j1}^-)^2 + (a_{ij2} - a_{j2}^-)^2 + (a_{ij3} - a_{j3}^-)^2 + (a_{ij4} - a_{j4}^-)^2}{4}} \end{cases} \tag{8.6}$$

It should be noted that the incorrect way may lead to the reversal of the alternative. From Table 8.3, one can see that alternative 4 is better than alternative 5 in the case of five alternatives, whereas reverses in the case of six alternatives. The reason is that in these two cases there exist different PIP and NIP.

In fact, the reversal is normal because the TOPSIS is only a relative method, whereas the DM may be confused. So, one can avoid a confusing result through minor modification of the selection of PIP and NIP, which will be chosen among all possible alternatives, including some potential alternatives not being evaluated. For example, if one needs to rank five alternatives, one can consider the potential comparable alternative 6 at the same time, which does not mean that

Table 8.3 The Reversal Result

	C_1	C_2	C_3	C_4	FIVE ALTERNATIVES (SCORE)	FIVE ALTERNATIVES (RANK)	SIX ALTERNATIVES (SCORE)	SIX ALTERNATIVES (RANK)
A_1	8	6.5	4.5	8.7	0.60	1	0.55	1
A_2	7.5	5.5	6.7	4.5	0.29	5	0.27	5
A_3	6.5	8	3.2	7.8	0.49	2	0.44	2
A_4	9	4.5	8.9	2.3	0.41	3	0.33	4
A_5	10	3.4	7.8	4.5	0.33	4	0.36	3
A_6	9	12	9	9				

Table 8.4 The Reversal Results of Different PIPs

	C_1	C_2	C_3	C_4	*	RANK	**	RANK	***	RANK
A_1	8	6.5	4.5	8.7	5.061	1	19,986.150	1	16.468	1
A_2	7.5	5.5	6.7	4.5	5.914	2	19,987.900	5	18.045	4
A_3	6.5	8	3.2	7.8	6.749	4	19,987.250	3	17.673	2
A_4	9	4.5	8.9	2.3	7.363	5	19,987.651	4	18.567	5
A_5	10	3.4	7.8	4.5	6.325	3	19,987.151	2	17.935	3
*	10	8	8.9	8.7						
**	10,000	10,000	10,000	10,000						
***	15	15	15	15						

* and ** indicate distances between the alternative and the PIP; *** indicates confusing results among PIPs.

one cannot select the "+∞" and "-∞" (the negative infinity and positive infinity). The selection of a positive or negative infinity should not change the rank result but should lessen the distinction among the alternatives. The selection will increase the difficulty of making a decision. In Table 8.4, we rank the alternatives A_1–A_5. From these experiments, we can conclude several rules about the choice of the PIP. First, the PIP and NIP should be the benchmark of the alternatives. Second, they should be realistic. An excellent alternative will not only affect the rank from the mathematical result, but it is not good for improving the relatively poor alternative. On the other hand, it may increase the difficulty of selecting a better alternative since it decreases the distinguishable degree among alternatives. Third, the ideal point should go beyond the existing alternatives. At least, the potential alternatives should be considered.

3. The closeness coefficient of alternative: A closeness coefficient is defined to determine the rank order of all candidates once $d(a_i, a^+)$ and $d(a_i, a^-)$ of each alternative are calculated, which can be defined as

$$d_i = \frac{d(a_i, a_j^-)}{d(a_i, a_j^-) + d(a_i, a_j^+)}, \; i = 1, \ldots, m \qquad (8.7)$$

One can rank the alternatives through the value of d_i. The greater d_i is, the better alternative i.

8.2.2 TOPSIS of Complete Weight with Multistage Linguistic Evaluation for Group Decision

Let a_{ij}^{kt} be the linguistic evaluation of DM k, $k = 1, \ldots, K$ in stage t, $t = 1, \ldots, T$, for criteria j. Assume that the weight of criteria is w_j^t, the weight of stage is w^t, and the weight of the DM is λ_k. Based on different aggregation order, two different methods are developed.

Method 1: First, calculate the distance between the alternatives and PIP and NIP of each decision maker. In this method, we first calculate the PIP and NIP of each decision maker. Then obtain the closeness coefficient via the weight of the decision maker and weight of stage.

Step 1: Set up the PIP and NIP for each stage.
Based on the note of selection of PIP and NIP, let the PIP and NIP in stage t be

$$\begin{cases} a_j^{t+} = (a_{j1}^{t+}, a_{j2}^{t+}, a_{j3}^{t+}, a_{j4}^{t+}), j = 1, \ldots, m, t = 1, \ldots, T \\ a_j^{t-} = (a_{j1}^{t-}, a_{j2}^{t-}, a_{j3}^{t-}, a_{j4}^{t-}), j = 1, \ldots, m, t = 1, \ldots, T \end{cases} \qquad (8.8)$$

Step 2: Calculate the distance between the alternative and the PIP and NIP.

First, the distance between the alternatives and PIP and NIP for DM k in stage t can be calculated as Formula (8.9), where the weight of criteria j can be obtained from Formula (8.2).

$$
\begin{cases}
d_i^{kt+} = d(a_i^{kt}, a^{t+}) = \sum_{j=1}^{n} w_j^t \\[2mm]
\sqrt{\dfrac{[(a_{ij1}^{k} - a_{j1}^{t+})^2 + (a_{ij2}^{k} - a_{j2}^{t+})^2 + (a_{ij3}^{k} - a_{j3}^{t+})^2 + (a_{ij4}^{k} - a_{j4}^{t+})^2]}{4}} \\[4mm]
d_i^{kt-} = d(a_i^{kt}, a^{t-}) = \sum_{j=1}^{n} w_j^t \\[2mm]
\sqrt{\dfrac{[(a_{ij1}^{k} - a_{j1}^{t-})^2 + (a_{ij2}^{k} - a_{j2}^{t-})^2 + (a_{ij3}^{k} - a_{j3}^{t-})^2 + (a_{ij4}^{k} - a_{j4}^{t-})^2]}{4}}
\end{cases}
$$

$$(8.9)$$

Second, the distance of the alternatives for all stages can be calculated as

$$
\begin{cases}
d_i^{k+} = \sum_{t=1}^{T} w^t d(a_i^{kt}, a^{t+}) \\[4mm]
d_i^{k-} = \sum_{t=1}^{T} w^t d(a_i^{kt}, a^{t-})
\end{cases}
$$

$$(8.10a)$$

or

$$
\begin{cases}
d_i^{t+} = \sum_{k=1}^{K} \lambda_k d(a_i^{kt}, a^{t+}) \\[4mm]
d_i^{t-} = \sum_{k=1}^{K} \lambda_k d(a_i^{kt}, a^{t-})
\end{cases}
$$

$$(8.10b)$$

Third, the distance of the alternatives for all of DMs can be calculated as

$$\begin{cases} d_i^+ = \sum_{k=1}^{K} \lambda_k \sum_{t=1}^{T} w^t d(a_i^{kt}, a^{t+}) \\ d_i^- = \sum_{k=1}^{K} \lambda_k \sum_{t=1}^{T} w^t d(a_i^{kt}, a^{t-}) \end{cases} \qquad (8.11a)$$

or

$$\begin{cases} d_i^+ = \sum_{t=1}^{T} w^t \sum_{k=1}^{K} \lambda_k d(a_i^{kt}, a^{t+}) \\ d_i^- = \sum_{t=1}^{T} w^t \sum_{k=1}^{K} \lambda_k d(a_i^{kt}, a^{t-}) \end{cases} \qquad (8.11b)$$

Step 3: Calculate the closeness coefficient.
The closeness coefficient can be calculated as

$$d_i^{m1} = \frac{d_i^-}{d_i^+ + d_i^-} = \frac{\sum_{k=1}^{K} \lambda_k \sum_{t=1}^{T} w^t d(a_i^{kt}, a^{t-})}{\sum_{k=1}^{K} \lambda_k \sum_{t=1}^{T} w^t d(a_i^{kt}, a^{t+}) + \sum_{k=1}^{K} \lambda_k \sum_{t=1}^{T} w^t d(a_i^{kt}, a^{t-})}$$

$$(8.12)$$

It is easy to see that the results from Formulas (8.10a), (8.11a) and (8.10b), (8.11b) are identical.

Method 2: First obtain the integration preference via the weight of the decision maker. In this method we first obtain the integration preference via the weight of the decision maker compared with method 1. Then, calculate the NIP and PIP for each stage. After that, we calculate the distance via the weight of the stage.

Step 1: With the weight of the DM, one can obtain the integration preference a_{ij}^t as

$$a_{ij}^t = \sum_{k=1}^{K} a_{ij}^{kt} \lambda_k \qquad (8.13)$$

In Formula (8.13), the operation rule of calculating fuzzy numbers should be used. Suppose that $m = (m_1, m_2, m_3, m_4)$ and $n = (n_1, n_2, n_3, n_4)$ are two trapezoidal fuzzy numbers. The addition operation can be calculated as Formula (8.14a), and the multiplication of any real number γ can be calculated as Formula (8.14b).

$$m + n = (m_1 + n_1, m_2 + n_2, m_3 + n_3, m_4 + n_4) \tag{8.14a}$$

$$m\gamma = (m_1\gamma, m_2\gamma, m_3\gamma, m_4\gamma), \gamma \geq 0 \tag{8.14b}$$

Step 2: One can calculate the distance between the alternatives and the PIP and NIP via TOPSIS, where the decision matrix is $(a_{ij}^t)_{m \times n}$. Suppose a^{t+} is (100, 100, 100, 100) and a^{t-} is (0, 0, 0, and 0). Then, one can obtain Formula (8.15).

$$\begin{cases} d_i^{t+} = d(a_i^t, a^{t+}) \\ d_i^{t-} = d(a_i^t, a^{t-}) \end{cases} \tag{8.15}$$

Step 3: Let the closeness coefficient be

$$d_i^{m2} = \frac{\sum_{t=1}^{T} w^t d_i^{t-}}{\sum_{t=1}^{T} w^t d_i^{t+} + \sum_{t=1}^{T} w^t d_i^{t-}}.$$

As a result, the final integration closeness coefficient from these two methods can also be aggregated as

$$d_i = \frac{d_i^{m1} + d_i^{m2}}{2}, i = 1, \ldots, m$$

Then we can rank the alternative through the value of d_i.

8.2.3 TOPSIS of Incomplete Weight of DM and Stage with Multistage Linguistic Evaluation for Group Decision

The rank of the alternatives is determined by the weight of the DM and the weight of stage. In many cases, it is difficult to give these weights exactly due to the uncertainty of the decision-making problem. In this part we will discuss how to rank the alternatives without complete weight information.

Method 3: Rank the alternative based on the weight model for both stage and DM.

In some decision-making cases, the difference among the alternatives should be clear; otherwise, it will be hard to distinguish. For example, suppose that the closeness coefficient of alternative 1 is 0.4546 and for alternative 2 it is 0.4547. Under this condition, we are not sure of the rank because these two coefficients are almost equal. Generally, Formula (8.16) should be satisfied in order to best differentiate the alternatives, where d_i^{m1} is as Formula (8.12).

$$\sum_{i=1}^{m} \sum_{j=1,j>i}^{m} (d_i^{m1} - d_j^{m1})^2 \rightarrow \max \tag{8.16}$$

For convenience, Formula (8.16) can be simplified as

$$\sum_{i=1}^{m} (d_i^{m1} - \frac{\sum_{i=1}^{m} d_i^{m1}}{m})^2 \rightarrow \max \tag{8.17}$$

Assume that there exists some incomplete prior information of the weight the of DM and the weight of stage, which can be expressed as Formula (8.18a) and Formula (8.18b), respectively. Formula (8.18a) means the information for the stage, and Formula (8.18b) means the information for the decision maker.

$$\begin{cases} w^t \geq w^s \\ w^t - w^s \geq \alpha_i \\ w^t \geq \alpha_i w^s \\ w^t - w^s \geq w^t - w^w \\ w^t \in [w^{tL}, w^{tU}] \end{cases} \quad (8.18a) \quad \text{or} \quad \begin{cases} \lambda_k \geq \lambda_p \\ \lambda_k - \lambda_p \geq \alpha_i \\ \lambda_k \geq \alpha_i \lambda_p \\ \lambda_k - \lambda_p \geq \lambda_k - \lambda_s \\ \lambda_k \in [\lambda_k^L, \lambda_k^U] \end{cases} \quad (8.18b)$$

Let H^t and H^k denote the weight set for stage and the decision maker. One model $P_{8.1}$ to estimate the weight for stage and decision maker can be suggested.

$$\max df = \sum_{i=1}^{m} \left(d_i^{m1} - \frac{\sum_{i=1}^{m} d_i^{m1}}{m} \right)^2$$

$$
\left\{
\begin{array}{l}
d_i^{m1} = \dfrac{\displaystyle\sum_{k=1}^{K} \lambda_k \sum_{t=1}^{T} w^t d(a_i^{kt}, a^{t-})}{\displaystyle\sum_{k=1}^{K} \lambda_k \sum_{t=1}^{T} w^t d(a_i^{kt}, a^{t+}) + \sum_{k=1}^{K} \lambda_k \sum_{t=1}^{T} w^t d(a_i^{kt}, a^{t-})}, \\[3em]
i = 1, \ldots, m \\[2em]
d(a_i^{kt}, a^{t+}) = \displaystyle\sum_{j=1}^{n} w_j^t \\[2em]
\sqrt{\dfrac{[(a_{ij1} - a_{j1}^{t+})^2 + (a_{ij2} - a_{j2}^{t+})^2 + (a_{ij3} - a_{j3}^{t+})^2 + (a_{ij4} - a_{j4}^{t+})^2]}{4}} \\[2em]
d(a_i^{kt}, a^{t-}) = \displaystyle\sum_{j=1}^{n} w_j^t \\[2em]
\sqrt{\dfrac{[(a_{ij1} - a_{j1}^{t-})^2 + (a_{ij2} - a_{j2}^{t-})^2 + (a_{ij3} - a_{j3}^{t-})^2 + (a_{ij4} - a_{j4}^{t-})^2]}{4}} \\[2em]
\displaystyle\sum_{k=1}^{K} \lambda_k = 1, \ \sum_{t=1}^{T} w^t = 1 \\[2em]
\lambda_k \in H^k, \ w^t \in H^t \\[1em]
\lambda_k \geq 0, \ w^t \geq 0, \ k = 1, \ldots, K, \ t = 1, \ldots, T
\end{array}
\right.
\qquad P_{8.1}
$$

The above method can be summarized in three steps.

Step 1: Analyze the weight of stage and decision maker and suppose that they are H.

Step 2: Calculate the weight of stages through model $P_{8.1}$.

Step 3: Calculate the distance between the alternatives and NIP and PIP via Formulas (8.10a), (8.11a) or (8.10b), (8.11b). Then calculate the closeness coefficient via Formula (8.12).

Method 4: Rank the alternative based on the weight model for stage.

One can obtain the weight of the DM and weight of stage through model $P_{8.1}$. In fact, we can also obtain the weight for the decision maker via the following method. In group decision making, the consistency of the group is often required. The similarity of the decision maker's preference is adopted to measure the group's consistency. The closer the decision maker to the group, the greater the weight for the decision maker is. So, before solving model $P_{8.1}$, we estimate the weight of the DM, which is considered from two angles. First, it is estimated by the Euclidean distance, which measures the difference as

$$
\lambda_k^E = \frac{\displaystyle\prod_{t=1}^{T}\prod_{i=1}^{n}\sum_{j=1}^{m}\left(a_{ij}^{kt} - \sum_{k=1}^{K} a_{ij}^{kt}/K\right)^2}{\displaystyle\sum_{k=1}^{K}\prod_{t=1}^{T}\prod_{i=1}^{n}\sum_{j=1}^{m}\left(a_{ij}^{kt} - \sum_{k=1}^{K} a_{ij}^{kt}/K\right)^2}
\tag{8.19}
$$

Formula (8.19) mainly considers the distance between the decision maker and the group. However, it cannot take the rank of the alternative into account. For example, suppose the preferences of three decision makers are respectively (60, 80, 70), (55, 72, 80), and (40, 50, 45). From Formula (8.19), the distance between decision maker 1 and decision maker 2 is 13.75, and the distance between 1 and 3 is 43.87. The former is smaller than the latter. However, we have to consider the rank of the alternatives, which is (3, 1, 2), (3, 2, 1), and (3, 1, 2). With respect to the rank of the alternative, the relationship between decision maker 1 and decision maker 3 is closer than 1 and 2. In this case, the method of gray relationship is adopted to determine the weight of the decision maker based on the closeness of the preference of group.

Definition 8.1

Suppose the sequences of system and behavior are [106, 107]

$$X_0 = [x_0(1), x_0(2), \dots, x_0(n)]$$
$$X_1 = [x_1(1), x_1(2), \dots, x_1(n)]$$

...   .

$$X_m = [x_m(1), x_m(2), \dots, x_m(n)]$$

Let $\quad \gamma[x_0(k), x_i(k)] = \dfrac{\min\limits_{i} \min\limits_{k} |x_0(k) - x_i(k)| + \xi \max\limits_{i} \max\limits_{k} |x_0(k) - x_i(k)|}{|x_0(k) - x_i(k)| + \xi \max\limits_{i} \max\limits_{k} |x_0(k) - x_i(k)|}$

$$(8.20)$$

$$\gamma[X_0, X_i] = \frac{1}{n} \sum_{k=1}^{n} \gamma[x_0(k), x_i(k)] \qquad (8.21)$$

where ξ is a parameter and $\xi \in (0, 1)$. $\gamma(X_0, X_i)$ is called the gray relationship between X_0 and X_i. It measures the closeness among the group via the similarities of the shapes of the sequences. For example, suppose four sequences are $X_0 = (45.8, 43.4, 42.3, 41.9)$, $X_1 = (39.1, 41.6, 43.9, 44.9)$, $X_2 = (34, 33, 35, 35)$, and $X_3 = (67, 68, 54, 47)$. Their shapes are pictured in Figure 8.1. The gray relationship of $\gamma_{01} = 0.55$, $\gamma_{02} = 0.71$, and $\gamma_{03} = 0.63$ can be obtained. The calculation process can be operated via software, which can be downloaded freely from http://igss. nuaa.edu.cn/institute. In addition, the gray relationship of (60, 80, 70),

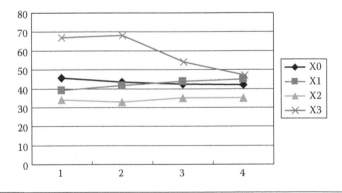

Figure 8.1　The shape of four sequences.

(55, 72, 80), and (40, 50, 45) is 0.7297 and 0.8029 according to the basis of the preference of decision maker 1.

Assume that the preference a_{ij}^{kt} is the sequence $X_1^t, X_2^t, \cdots, X_k^t$ in our problem. Let the average preference of the group be $X_0^t = \Sigma_{k=1}^{K} a_{ij}^{kt}/K$. As a result, the weight of the DM can be defined as

$$\lambda_k^G = \frac{\prod\limits_{t=1}^{T} \sqrt[m]{\prod\limits_{j=1}^{m} \gamma(X_{0j}^t, X_{kj}^t)}}{\sum\limits_{k=1}^{K} \prod\limits_{t=1}^{T} \sqrt[m]{\prod\limits_{j=1}^{m} \gamma(X_{0j}^t, X_{kj}^t)}}, \quad k = 1, \ldots, K \qquad (8.22)$$

With respect to the two results, the weight of the decision maker can be defined as

$$\lambda_k = \frac{\lambda_k^E \lambda_k^G}{\sum\limits_{k=1}^{K} \lambda_k^E \lambda_k^G}, \quad k = 1, \ldots, K \qquad (8.23)$$

Through λ_k, we can obtain the weight of stage via model $P_{8.1}$. Then, we can calculate d_i^{m1} and rank the alternatives. The calculation process can be concluded as follows:

Step 1: Calculate of the weight of DMs according to Formulas (8.19)–(8.23).

Step 2: Calculate of the weight of stages through model $P_{8.1}$.

Step 3: Calculate the NIP and PIP via Formulas (8.10a), (8.11a) or (8.10b), (8.11b). Then calculate the closeness coefficient through Formula (8.12).

8.3 Example Analysis

8.3.1 Problem and Basic Information

Assume that one equipment vendor can be selected via four criteria: advanced technology (C_1), maturity (C_2), guarantee after sale (C_3), and price for production and service (C_4). Suppose that five vendors are to be selected, which are denoted as a_1–a_5. In order to understand the complete information of the vendors, the production and technology information of vendors among three stages is obtained by five DMs (Table 8.5).

Table 8.5   Linguistic Ratings for the Alternatives

STAGE 1	ALT	C1	C2	C3	C4	STAGE 2	ALT	C1	C2	C3	C4	STAGE 3	ALT	C1	C2	C3	C4
DM 1	a1	s8	s2	s4	s6	DM 1	a1	s6	s8	s2	s4	DM 1	a1	s3	s4	s7	s5
	a2	s3	s6	s8	s4		a2	s4	s6	s7	s5		a2	s5	s3	s8	s5
	a3	s3	s5	s8	s5		a3	s5	s6	s2	s1		a3	s8	s5	s5	s3
	a4	s7	s3	s7	s7		a4	s3	s6	s8	s2		a4	s5	s3	s3	s5
	a5	s3	s8	s8	s5		a5	s7	s8	s5	s9		a5	s3	s4	s3	s8
DM 2	a1	s5	s6	s7	s3	DM 2	a1	s3	s6	s5	s5	DM 2	a1	s3	s5	s4	s9
	a2	s8	s6	s4	s9		a2	s5	s5	s2	s6		a2	s3	s3	s1	s8
	a3	s2	s5	s6	s7		a3	s1	s6	s3	s5		a3	s5	s6	s3	s5
	a4	s7	s8	s3	s6		a4	s4	s6	s5	s7		a4	s3	s3	s5	s3
	a5	s1	s2	s8	s7		a5	s3	s6	s7	s8		a5	s3	s4	s5	s8
DM 3	a1	s2	s8	s4	s5	DM 3	a1	s3	s5	s6	s3	DM 3	a1	s5	s3	s4	s7
	a2	s3	s4	s8	s6		a2	s4	s6	s8	s8		a2	s6	s4	s6	s4
	a3	s6	s6	s3	s8		a3	s4	s2	s6	s5		a3	s5	s7	s3	s6
	a4	s8	s4	s5	s6		a4	s6	s6	s8	s3		a4	s3	s8	s5	s4
	a5	s6	s7	s7	s6		a5	s7	s8	s5	s6		a5	s5	s4	s6	s8
DM 4	a1	s3	s5	s7	s8	DM 4	a1	s6	s8	s9	s4	DM 4	a1	s5	s3	s2	s7
	a2	s3	s6	s3	s7		a2	s3	s5	s6	s5		a2	s4	s5	s7	s4
	a3	s5	s8	s5	s5		a3	s5	s7	s5	s8		a3	s5	s2	s5	s8
	a4	s4	s7	s3	s6		a4	s2	s4	s7	s8		a4	s3	s7	s5	s5
	a5	s2	s6	s7	s8		a5	s5	s7	s8	s5		a5	s4	s7	s5	s5
DM 5	a1	s8	s1	s5	s7	DM 5	a1	s4	s6	s1	s9	DM 5	a1	s3	s8	s6	s4
	a2	s3	s3	s7	s8		a2	s3	s7	s8	s5		a2	s4	s5	s7	s3
	a3	s3	s1	s6	s9		a3	s4	s8	s9	s6		a3	s3	s6	s5	s2
	a4	s2	s5	s4	s6		a4	s5	s8	s5	s5		a4	s3	s6	s8	s5
	a5	s3	s6	s4	s8		a5	s4	s6	s2	s5		a4	s7	s3	s5	s9

Table 8.6   The Linguistic Importance and Weights of DMs

	DM 1	DM 2	DM 3	DM 4	DM 5
Linguistic importance for DMs	VH	M	M	L	M
	9.25	5	5	2.75	5
Weights of DMs	0.343	0.185	0.185	0.102	0.185

Table 8.7   The Weights of Criteria

	STAGE 1			STAGE 2			STAGE 3		
C1	VH	9.25	0.381	C1	H	7.25	0.274	C1	H	7.25	0.252
C2	H	7.25	0.299	C2	M	5	0.189	C2	H	7.25	0.252
C3	L	2.75	0.113	C3	VH	9.25	0.349	C3	VH	9.25	0.322
C4	M	5	0.206	C4	M	5	0.189	C4	M	5	0.174

Table 8.8 The Weights of Stage

STAGE 1	STAGE 2	STAGE 3
M	H	VH
5	7.25	9.25
0.233	0.337	0.430

8.3.2 TOPSIS of Complete Weight for Criteria and DMs and Stages

- The weights of DM are shown in Table 8.6
- The weights of criteria are shown in Table 8.7
- The weights of stage are shown in Table 8.8

1. TOPSIS based on Method 1

Step 1: Suppose that the PIP is S9 and the NIP is S1 for each criterion (S9 and S1 are linguistic values shown in Table 8.1). That is,

$$\begin{cases} a_j^+ = (a_{j1}^+, a_{j2}^+, a_{j3}^+, a_{j4}^+) = (80, 100, 100, 100) \\ a_j^- = (a_{j1}^-, a_{j2}^-, a_{j3}^-, a_{j4}^-) = (0, 0, 0, 20) \end{cases}$$

Step 2: Calculate the distance between the alternatives and the PIP and NIP for each criterion. The result is shown in Table 8.9.

Then calculate the distance between the alternatives and PIP and NIP considering the weight of stage (results of columns 2 and 4 in Table 8.10) and the weight of the decision maker (columns 3 and 5 column in Table 8.10).

Step 3: Calculate the closeness coefficient. The result is shown in Table 8.11.

2. TOPSIS based on Method 2

Step 1: Obtain the integration preference expressed by the trapezoidal fuzzy numbers with the weight of the decision maker (Table 8.12) for each stage.

Step 2: Calculate the distance between the alternatives and the PIP and NIP for each criterion (Table 8.13). The distance considers the weight of the criteria as in Table 8.14.

Step 3: Calculate the closeness coefficient and obtain the rank of the alternative (Table 8.15). We can see that the rank of alternative 2 and alternative 3 reverses.

Table 8.9   The Distance between Alternatives and PIP and NIP for Each Criterion

PIP	STAGE 3	STAGE 2	STAGE 1	NIP	STAGE 3	STAGE 2	STAGE 1
DM 1	48.629	53.085	44.475	DM 1	40.996	44.608	52.505
	43.115	39.036	54.508		54.929	57.059	42.422
	42.996	64.913	53.551		49.799	29.481	38.969
	62.855	47.853	34.714		41.147	47.772	58.105
	61.837	23.403	44.079		43.108	69.065	49.557
DM 2	51.070	52.230	45.996	DM 2	52.424	41.388	48.639
	69.672	56.474	23.596		41.332	38.127	72.852
	53.082	66.706	52.427		25.098	26.700	41.878
	65.952	43.607	26.972		22.376	52.300	68.172
	52.290	36.302	64.005		33.745	58.428	29.008
DM 3	54.082	56.372	51.882	DM 3	37.264	38.867	43.298
	48.203	31.948	57.004		43.922	65.745	39.926
	46.605	54.211	36.478		29.457	43.581	62.016
	48.417	36.166	37.193		27.015	61.044	60.292
	41.885	30.335	28.733		47.925	63.987	68.530
DM 4	59.745	24.728	47.379	DM 4	31.254	71.280	45.620
	44.245	50.773	51.916		42.216	44.245	43.164
	49.246	34.469	36.353		43.995	58.106	56.414
	46.120	42.948	45.591		30.648	52.483	51.673
	43.130	29.388	46.854		35.545	63.786	49.733
DM 5	45.494	55.952	41.235	DM 5	34.799	39.145	51.752
	46.307	37.506	56.250		40.135	55.987	37.097
	57.177	27.012	60.038		25.138	68.996	32.495
	40.824	39.848	59.482		42.042	52.508	36.383
	38.150	61.347	49.916		53.097	35.642	46.550

8.3.3 TOPSIS of Weight for Criteria and DMs and Stages with Incomplete Information

1. TOPSIS based on Method 3

Step 1: Suppose that the prior information of the weight of stage and the weight of decision maker is as

$$
\begin{cases}
w^1 \in (0.1,\ 0.2) \\
w^2 \in (0.2,\ 0.35); \\
w^3 \geq w^2
\end{cases}
\begin{cases}
\lambda_k \geq 0.1,\ k = 1, \ldots, 5 \\
\lambda_2 \in (0.3,\ 0.4) \\
\lambda_3 \geq \lambda_4
\end{cases}
\tag{8.24}
$$

Step 2: According to Formula (8.24), the weight of stage and the weight of the decision maker (Table 8.16) can be obtained via model $P_{8.1}$.

Table 8.10 The Distance between Alternative and PIP and NIP Considering the Weights of Stages and DMs

DM	PIP		NIP	
DM	WITH STAGE WEIGHT	WITH DM WEIGHT	WITH STAGE WEIGHT	WITH DM WEIGHT
DM 1	48.345	16.582	47.166	16.178
	46.641	15.998	50.266	17.241
	54.928	18.840	38.288	13.133
	45.689	15.671	50.677	17.382
	41.236	14.144	54.635	18.740
DM 2	49.278	9.116	47.074	8.709
	45.398	8.399	53.812	9.955
	57.394	10.618	32.858	6.079
	41.647	7.705	52.170	9.651
	51.939	9.609	40.030	7.406
DM 3	53.908	9.973	40.401	7.474
	46.508	8.604	49.562	9.169
	44.812	8.290	48.228	8.922
	39.457	7.300	52.807	9.769
	32.332	5.981	62.206	11.508
DM 4	42.617	4.347	50.932	5.195
	49.747	5.074	43.308	4.417
	38.716	3.949	54.096	5.518
	44.823	4.572	47.056	4.800
	40.098	4.090	51.172	5.220
DM 5	47.188	8.730	43.558	8.058
	47.617	8.809	44.174	8.172
	48.236	8.924	43.092	7.972
	48.522	8.977	43.137	7.980
	51.035	9.441	44.395	8.213

Table 8.11 The NIP and PIP

	ALT 1	ALT 2	ALT 3	ALT 4	ALT 5
NIP	45.614	48.955	41.624	49.583	51.086
PIP	48.748	46.884	50.621	44.224	43.266
Closeness coefficient	0.483	0.511	0.451	0.529	0.541
Rank	4	3	5	2	1

Table 8.12 The Integration Preference for Each Stage

STAGE 1	C1				C2				C3				C4			
	A	B	C	D	A	B	C	D	A	B	C	D	A	B	C	D
A1	37.23	53.53	67.79	77.23	19.71	29.15	45.26	61.56	22.77	42.77	58.61	78.61	33.06	53.06	65.39	83.35
A2	11.1	31.1	34.8	51.1	18.9	38.9	63.35	83.35	42.78	62.78	78.89	88.33	37.57	57.57	77.11	89.71
A3	8.61	24.91	34.16	54.16	27.51	43.81	51.4	69.36	34.74	54.74	72.7	85.84	50.35	70.35	74.05	86.65
A4	42.78	59.08	69.54	85.84	22.77	42.77	52.02	68.32	26.13	46.13	51.68	71.68	40.29	60.29	80	100
A5	5.55	19.81	27.4	47.4	40.29	56.59	75.76	88.9	48.9	68.9	85.01	94.45	44.16	64.16	75.45	89.71

STAGE 2	C1				C2				C3				C4			
	A	B	C	D	A	B	C	D	A	B	C	D	A	B	C	D
A1	13.35	33.35	52.25	72.25	43.35	63.35	83.35	94.45	19.26	28.7	41.11	59.07	20.35	40.35	53.7	70
A2	5.55	25.55	41.39	61.39	35.55	55.55	71.39	91.39	45.84	62.14	76.3	88.9	35.55	55.55	64.8	81.1
A3	13.35	29.65	40.75	60.75	33.06	49.36	72.6	88.9	23.41	36.55	48.96	65.26	22.77	35.91	43.5	61.46
A4	11.1	29.06	42.2	62.2	32.49	52.49	80.64	96.94	48.9	68.9	79.46	88.9	22.77	35.91	44.81	62.77
A5	34.74	54.74	60.29	80.29	48.9	68.9	90.56	100	33.06	49.36	55.1	73.06	52.7	72.7	81.95	91.39

STAGE 3	C1				C2				C3				C4			
	A	B	C	D	A	B	C	D	A	B	C	D	A	B	C	D
A1	8.61	28.61	28.61	48.61	16.65	36.65	50.64	66.94	26.13	44.09	62.78	82.78	42.31	62.31	67.86	84.16
A2	15.84	35.84	50	70	8.61	28.61	34.16	54.16	43.35	59.65	72.06	85.2	21.39	41.39	53.7	70
A3	34.74	54.74	61.6	74.74	32.49	50.45	63.59	83.59	18.9	38.9	38.9	58.9	17.22	33.52	44.81	62.77
A4	10.29	30.29	30.29	50.29	22.77	42.77	52.02	68.32	25.26	45.26	48.96	65.26	18.9	38.9	44.45	64.45
A5	16.65	36.65	39.71	59.71	6.12	26.12	47.51	67.51	19.71	39.71	45.26	65.26	60.64	80.64	94.9	96.94

Table 8.13 The Integration Preference for Each Stage

STAGE 1: PIP	C1	C2	C3	C4	NIP	C1	C2	C3	C4
A1	43.748	63.127	53.410	45.165	a1	60.851	42.060	54.686	61.505
A2	69.449	54.614	36.198	39.773	a2	35.045	56.637	70.351	68.410
A3	71.461	54.113	42.583	32.388	a3	34.622	50.321	64.918	71.547
A4	38.977	56.003	53.609	37.223	a4	66.191	49.299	51.526	73.584
A5	76.467	39.244	30.959	35.746	a5	29.244	67.949	76.299	70.369
STAGE 2: PIP	C1	C2	C3	C4	NIP	C1	C2	C3	C4
A1	61.238	34.859	64.703	56.890	a1	48.064	73.757	39.918	49.563
A2	69.623	41.899	35.537	43.948	a2	39.260	66.705	70.156	61.494
A3	66.160	44.495	58.532	60.724	a3	40.026	64.621	46.206	43.237
A4	66.531	42.423	32.107	60.216	a4	40.674	70.198	73.068	44.034
A5	45.477	30.280	49.465	29.080	a5	59.759	79.592	54.551	76.044
STAGE 3: PIP	C1	C2	C3	C4	NIP	C1	C2	C3	C4
A1	72.777	60.187	50.657	38.835	a1	31.914	46.545	57.924	65.880
A2	60.414	70.507	38.200	56.255	a2	47.264	35.330	66.875	49.886
A3	45.880	46.389	62.715	62.658	a3	58.275	60.481	41.391	42.919
A4	71.130	56.003	55.665	60.539	a4	33.429	49.299	48.328	44.721
A5	63.676	67.236	59.759	22.133	a5	41.118	43.401	45.477	84.533

Table 8.14 The Distance between the Alternatives and PIP and NIP

	STAGE 1		STAGE 2		STAGE 3	
	PIP	NIP	PIP	NIP	PIP	NIP
A1	50.88227	54.60995	56.70118	50.40856	56.57579	49.88629
A2	55.07344	52.32875	47.70447	59.47138	55.08085	51.02749
A3	54.89026	50.31153	58.44182	47.47807	54.34868	50.7223
A4	45.32099	60.93963	48.83379	58.23503	60.49564	44.19056
A5	51.73013	54.57646	40.94304	64.8274	56.08332	50.65104

Table 8.15 The Closeness Coefficient

	ALT 1	ALT 2	ALT 3	ALT 4	ALT 5
NIP	51.163	54.176	49.533	52.826	56.343
PIP	55.291	52.593	55.854	53.030	49.967
Closeness coefficient	0.481	0.507	0.470	0.499	0.530
Rank	4	2	5	3	1
Closeness coefficient (method 1)	0.483	0.511	0.451	0.529	0.541
Rank (method 1)	4	3	5	2	1

Step 3: Obtain the closeness coefficient (Table 8.17).

2. TOPSIS based on Method 4

Step 1: Estimate the weights of DMs.
First, calculate the weight of the DM using the Euclidean distance (Table 8.18).
Second, we calculate the weight of the DM from the method of gray relationship. The gray relationship of the decision maker for each stage is shown in Table 8.19. The final weight of the decision maker from the gray relationship is shown in Table 8.20.
Third, the final integration weight of the decision maker is shown in Table 8.21.
Step 2: Suppose that the weight of stage is as

$$\begin{cases} w^1 \in (0.1, 0.2) \\ w^2 \in (0.2, 0.35) \\ w^3 \geq w^2 \end{cases} \tag{8.25}$$

Then, we can obtain the weight of the stage; that is, 0.1, 0.35, and 0.55.
Step 3: The closeness coefficient is shown in Table 8.22.
The comparison result can be seen in Figures 8.2 and 8.3.

Table 8.16   The Weight of Stage and Decision Maker

STAGE	WEIGHT OF STAGE	DECISION MAKER	WEIGHT OF DECISION MAKER
w^1	0.1	λ_1	0.1
w^2	0.2	λ_2	0.3
w^3	0.7	λ_3	0.4
		λ_4	0.1
		λ_5	0.1

Table 8.17   The Closeness Coefficient

	ALT 1	ALT 2	ALT 3	ALT 4	ALT 5
NIP	45.657	50.130	46.452	55.394	51.731
PIP	49.056	46.149	47.538	39.398	42.908
Integration	0.482	0.521	0.494	0.584	0.547
Rank	5	3	4	1	2

Table 8.18 The Weight of DM Based on Euclidean Distance

	STAGE 1	STAGE 2	STAGE 3	WEIGHT OF DM
DM1	0.212	0.187	0.208	0.203
DM2	0.219	0.178	0.213	0.203
DM3	0.179	0.207	0.149	0.178
DM4	0.184	0.216	0.191	0.197
DM5	0.206	0.212	0.239	0.219

Table 8.19 The Gray Relationship of Decision Maker

		STAGE 1		STAGE 2		STAGE 3
CRITERIA	DM	GRAY RELATIONSHIP	DM	GRAY RELATIONSHIP	DM	GRAY RELATIONSHIP
C1	DM1	0.8992	DM1	0.8881	DM1	0.6281
	DM2	0.8460	DM2	0.7138	DM2	0.8272
	DM3	0.5701	DM3	0.5767	DM3	0.8556
	DM4	0.8209	DM4	0.8144	DM4	0.7728
	DM5	0.8561	DM5	0.8622	DM5	0.7303
C2	DM1	0.7860	DM1	0.7822	DM1	0.8886
	DM2	0.9463	DM2	0.7980	DM2	0.8494
	DM3	0.9254	DM3	0.5075	DM3	0.5963
	DM4	0.9617	DM4	0.6842	DM4	0.5793
	DM5	0.6747	DM5	0.6207	DM5	0.8021
C3	DM1	0.4928	DM1	0.8365	DM1	0.8382
	DM2	0.7430	DM2	0.9445	DM2	0.8389
	DM3	0.5942	DM3	0.9816	DM3	0.8506
	DM4	0.6962	DM4	0.9522	DM4	0.5415
	DM5	0.7720	DM5	0.6094	DM5	0.9405
C4	DM1	0.8560	DM1	0.7528	DM1	0.7348
	DM2	0.5134	DM2	0.9061	DM2	0.7970
	DM3	0.9192	DM3	0.6519	DM3	0.8664
	DM4	0.8671	DM4	0.7856	DM4	0.8006
	DM5	0.9376	DM5	0.7598	DM5	0.6912

Table 8.20 The Weight of Decision Maker Based on Gray Relationship

	STAGE 1	STAGE 2	STAGE 3	WEIGHT OF DM
DM1	0.7389	0.8133	0.7657	0.202
DM2	0.7434	0.8356	0.8279	0.210
DM3	0.7327	0.6578	0.7831	0.189
DM4	0.8309	0.8035	0.6637	0.199
DM5	0.8041	0.7055	0.7855	0.200

8.4 Summary and Future Work

Four methods are suggested to solve the decision problems of linguistics in multistages and multipersons. In the complete information of weight case, two different aggregations are developed. In the incomplete information case, the weight model is suggested. Also, two aggregations are put forward. One is to solve the incomplete information weight of stage and decision maker. The other one is to estimate

Table 8.21   Final Weight of Decision Maker

	WEIGHT OF DM (EUCLIDEAN DISTANCE)	WEIGHT OF DM (GRAY RELATIONSHIP)	FINAL WEIGHT
DM1	0.203	0.202	0.202
DM2	0.203	0.21	0.206
DM3	0.178	0.189	0.183
DM4	0.197	0.199	0.198
DM5	0.219	0.2	0.209

Table 8.22   The Closeness Coefficient

	ALT 1	ALT 2	ALT 3	ALT 4	ALT 5
NIP	46.986	48.547	44.528	51.863	50.935
PIP	47.370	46.759	48.642	42.473	43.366
Closeness coefficient	0.498	0.509	0.478	0.550	0.540
Rank	4	3	5	1	2

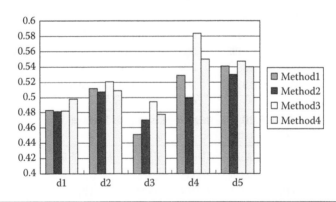

Figure 8.2   Comparison of closeness coefficient using four methods.

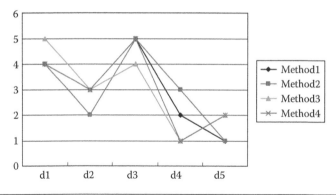

Figure 8.3 Comparison of rank of alternative using four methods.

the weight of the decision maker before solving the model. From comparison of the four methods, there exists much difference in the results. In the future, we will study which is suitable for the particular decision-making case. In addition, the weight model of stage and decision maker may be modified due to the changeable decision-making environment.

Conclusion and Summary

The common preference structures are judgment matrix, utility value, preference ordering value, linguistic value, and so on. Because of the uncertainty of the decision-making environment, decision makers tend to use expressions in the form of preference, such as interval numbers, fuzzy numbers, and linguistic variables. This book studies mainly group decision-making methods and group aggregation methods based on uncertainty preference information. The main content is as follows:

1. The decision-making method of uncertain method in group decision making: This book put forward a new uncertain judgment matrix—the three-point interval number judgment matrix. This book also researched its consistency and the method of solving weight. Based on the consistency concept of the three-point interval number judgment matrix, this book defined the most likely value deviation variables and upper and lower limitation variables and established the model of solving the weight of the three-point interval number judgment matrix.

2. Aggregation method of uncertain preference information in group decision making: The aggregation method of the interval number reciprocal judgment matrix and interval

number complementary judgment matrix was studied. This book used uncertain ordered weighted averaging (UOWA) to aggregate the preference information of decision makers to two forms—interval number reciprocal judgment matrix and interval number complementary judgment matrix, and then defined the satisfaction subordinate function and established the model in order to make sure the consistency degree of group preference is maximum.

Research was presented on the aggregation method of uncertain preference information of interval number preference sequence, interval number utility values, interval number reciprocal judgment matrix, and interval number complementary judgment matrix. It defined the consistency of the four uncertain preferences information and established the solution models of uncertain preference information, which have the same decision-making principles. The goal programming method of group ideas aggregated was put forward and research on the characteristics of this model was discussed.

Research was presented on the aggregation method of the three-point interval number reciprocal judgment matrix and the interval number complementary judgment matrix. Ordered weighted averaging (OWA) was used to aggregate preference information of decision makers to two three-point judgment matrix. The solution model of group preference according to the greatest consistency of decision makers' opinion was constructed.

3. The aggregation method of uncertain preference information with timing characteristics: Research was presented on the aggregation method based on multiple stages and structure form in the group decision-making process. Multiple preferences were made to be consistent according to the transform function of reciprocal judgment preference and complementary judgment preference. The method of certainty of the decision makers' weight based on the consistency level of decision makers and the deviation between decision makers' preference and group comprehensive preference was put

forward. The weight model based on multiple stages was established. The multiple-stage preference to group comprehensive preference according to the weight of each stage was aggregated.

Research was presented on the aggregation method of uncertain judge preference based on the multiple-stage and multiple-structure form. The united order model of multiple uncertain preference based on the definition of multiple uncertain preference was established. According to the deviation between individual preference and group preference, the method of certainty of the weight of decision-makers based on the gray correlation degree and Spearman's rank correlation coefficient method was put forward. The weight model based on multiple stages was established. We aggregated the multiple-stage preference to group comprehensive preference according to the weight of each stage.

In the group decision-making process, the following aspects should be considered in future research:

1. Dynamic disturbance of internal and external decision environment: Because of different benefits, there exist both a competition and cooperation relationship in the internal decision-making group and different groups, and so the decision-making process must be influenced by the internal and external environment. These complex factors may influence the preference form of decision makers, the aggregation method, the parameter of preference, and even the judge standard of decision making. These factors may change, so when one aggregates the group preference, the dynamic disturbance must be considered, and a comprehensive evaluation of these factors must be made, modifying the decision-making process and result in time.

2. The organizational structure complexity of decision-making groups and decision objects: Due to the deviation among personal prestige, political position, title, and administrative duties, decision makers make different decisions according to the specific decision objective. In some conditions, there exist

stratified characteristics and even complex relationships in the decision object because of the complex relationships in the alternatives. The organizational structure of the decision-making group and object may change when the decision-making environment changes; therefore, the change of organizational structure will influence the understanding and preference structure form of decision makers directly.

References

[1] Simon, H.A. 1955. A behaviour model of rational choice. *Quarterly Journal of Economics* 69:99–188.

[1a] Zhou, X., H. Zhenjie, L. Jing. 2004. Stochastic Borda number method for group decision making with stochastic preference. *OR Transactions* 8 (4):54–60.

[2] Hu, Y.-d. 2002. Group decision making with stochastic preference and impossibility theorem. *Progress in Natural Science* 12 (6):580–84.

[3] Hahn, E. D. 2006. Link function selection in stochastic multi-criteria decision making models. *European Journal of Operational Research* 172:86–100.

[4] Sabbadin, R. 2001. Possibilistic Markov decision processes. *Engineering Applications of Artificial Intelligence* 14:287–300.

[5] Hryniewicz, O. 2006. Possibilistic decisions and fuzzy statistical tests. *Fuzzy Sets and Systems* 157:2665–73.

[5a] Zadeh, L. 1965. Fuzzy sets. *Journal of Information and Control* 8:338–353.

[5b] Zadeh, L. 1978. Fuzzy sets as a basis for a theory of possibility. *Journal of Fuzzy Sets and Systems* 1:3–28.

[6] Gu, X., and Q. Zhu. 2006. Fuzzy multi-attribute decision-making method based on eigenvector of fuzzy attribute evaluation space. *Decision Support Systems* 41:400–410.

[7] Tang, J.-f., Y. Dong, S.-xin Liu, and D.-w. Wang. 2003. Aggregate production planning with fuzzy demands and fuzzy capacity. *Control Theory & Applications* 20 (6):894–96.

[8] Walle, B., and A. Rutkowski. 2006. A fuzzy decision support system for IT service continuity threat assessment. *Decision Support Systems* 42:1931–43.

[9] Li, Rongjun. 2002. *Fuzzy multiple criteria decision making and application.* Beijing: Science Press.

[10] Liu, X.-W., and S.-L. Han. 2005. Ranking fuzzy numbers with preference weighting function expectations. *Computers & Mathematics with Applications* 49 (11–12):1731–53.

[11] Deng, Y., Z. Zhenfu, and L. Qi. 2006. Ranking fuzzy numbers with an area method using radius of gyration. *Computers & Mathematics with Applications* 51 (6–7):1127–36.

[12] Deng, J.-l. 1987. *Primary methods of grey theory.* Wuhan: Huazhong University of Science and Technology (HUST) Press.

[13] Liu, S.-f., Y.-g. Dang, and Z.-g. Fang. 2004. *Methods and application of grey theory.* Beijing: Science Press.

[14] Liu, S.-f. 2004. Emergence and development of grey system theory and its forward trends. *Journal of Nanjing University of Aeronautics & Astronautics* 36 (2):267–72.

[15] Wang, G. 1990. Unascertained information and its mathematical treatment. *Journal of Harbin University of Civil Engineering and Architecture* 23 (4):1–9.

[16] Liu, K., H. Wu, and Y. Pang. 1999. *Uncertainty method and its application.* Beijing: Science Press.

[17] Parkan, C., and L.-F. Wang. 1997. Decision making under partial probability information. *Systems Engineering—Theory & Practice* 1 (2):23–32.

[18] Parkan, C., L.-F. Wang, and M.-L. Wu. 1999. Decision making under partial probability information using pair-wise comparisons. *European Journal of Operational Research* 112 (1):220–35.

[19] Yang, S.-l., Y.-z. Liu, and Y.-f. Li. Research on rough sets-based evidence acquirement and combination of DST. *Journal of Management Sciences in China* 8 (5):69–75.

[20] Lu, W.-h., and Q.-z. Wu. 2005. An expertise aggregation method based on the D-S evidence theory in group forecasting process. *Forecasting* 24 (4):66–69.

[21] Sikder, U., and A. Gangopadhyay. 2007. Managing uncertainty in location services using rough set and evidence theory. *Expert Systems with Applications* 32 (2):386–96.

[22] Qiu, W. 2002. *Management decision and entropy.* Beijing: Mechanism Industry Press.

[23] Chang, S.-L., R.-C. Wang, and S.-Y. Wang. 2007. Applying a direct multi-granularity linguistic and strategy-oriented aggregation approach on the assessment of supply performance. *European Journal of Operational Research* 177 (2):1013–25.

[24] Zhao, K. 2000. *Collect analysis and its application.* Zhejiang: Zhejiang Science Technology Press.

[25] Pawlak, Z. 1995. Vagueness and uncertainty: A rough set perspective. *Computational Intelligence* 11 (2):227–32.

[26] Wang, Q.-y., Y.-m. Cui, and B. Ren. 2000. The root of uncertainty information and the foundation of universal grey sets. *Journal of Huazhong University of Science and Technology* 28 (4):66–68.

[27] Liu, R.-x., D.-h. Wu, and H.-l. Ling. 2005. New research into judgment and decision-making under uncertainty—Prospect theory. *Operations Research and Management Science* 14 (2):14–18.

[28] Arrow, K. J. 1963. *Social choice and individual values.* New Haven, CT: Yale University Press.

[29] Bacharach, M. 1975. Group decisions in the face of difference of opinion. *Management Science* 22:2.

[30] Keeney, R., and C. Kirkwood. 1975. Group decision making using cardinal social welfare functions. *Management Science* 22 (44):430–37.

[31] Wang, X.-j. 2005. *Study of theory and method on some problems in group decision-making.* Dalian: Dongbei University.

[32] Zhou, X. 2004. *Some theories and methods in group decision-making and multiobjective decision-making.* Shanghai: Shanghai University.

[33] He, G. 2002. *Adaptive behavior and implicit learning mechanism in dynamic group decision making.* Hangzhou: Zhejiang University.

[34] Tung-K. 2006. Preference aggregation and conflict modeling in multi-attribute decision making. Ph.D. diss., State University of New York at Buffalo.

[35] Chen, Z. 2005. Consensus in group decision making under linguistic assessments. Ph.D. diss., Kansas State University.

[36] Xu, X. 2005. *Research of model-driven complex-large group decision support system in environment of network.* Wuhan: Zhongnan University.

[37] Pasi, G., and R. R. Yager. 2006. Modeling the concept of majority opinion in group decision making. *Information Sciences* 176:390–414.

[38] Li, W., Y. m. Xi, and S. w. Cheng. 2002. Review of process organizing research of group decision making. *Journal of Management Sciences in China* 5 (2):55–66.

[39] Wei, C.-p., and W.-h. Qiu. 2000. A review of the theory of group decision making. *Journal of Beijing University of Aeronautics and Astronautics (Social Sciences Edition)* 13 (2):24–28.

[40] Chen, X., and W. Bi. 2005. On multiagent framework of group decision support system generator. *Computer Engineering and Applications* 34:92–94.

[41] Chen, X.-h., and X.-h. Xu. 2006. Development of development platform of applied software developer kit for decision-smart decision. *Chinese Journal of Management* 3 (3):253–57.

[42] Dai, L. W. 1999. *Some problems about complexity science.* Beijing: Democracy and Construction Press.

[43] Yu, J.-y. 1993. *Study on the complexity.* Beijing: Science Press.

[44] Dai, R.-w. 2004. The research on systems complexity—Long-term and huge task. *Complex Systems and Complexity Science* 1 (3):1–3.

[45] Qian, X. 1990. A new discipline of science—The study of open complex giant system and its methodology. *Chinese Journal of Nature* 1:3–10.

[46] Gu, J.-f., and X.-j. Tang. 2005. Metasynthesis approach: Theory and applications. *Journal of Systemic Dialectics* 13 (4):1–7.

[47] Beynon, M., B. Curry, and P. Morgan. 2000. The Dempster–Shafer theory of evidence: An alternative approach to multicriteria decision modeling. *Omega* 2 (1):37–50.

[48] Wang, X.-r. 2003. Approach to multiple attribute group decision making with linguistic assessment information. *Journal of Systems Engineering* 8 (2):173–76.

[49] Wang, J.-q. 2003. The optimal assignment method for group decision making of multi-decision-maker. *Journal of Central South University (Science and Technology)* 34 (5):584–86.

[50] Xu, Z.-s. 2005. Interactive approach based on incomplete complementary judgement matrices to group decision making. *Control and Decision* 20 (8):913–16.

[51] Sun, J., W.-s. Xu, and Q.-d. Wu. 2006. A new algorithm for incomplete matrixes' compatibility improvement and group ranking in group decision making. *Systems Engineering—Theory & Practice* (10):88–94.

[52] Dong, Y.-c., Y.-h. Chen, and S. Wang. 2004. Algorithm of solving weights for group decision making by improving compatibility. *Systems Engineering —Theory & Practice* 10:86–91.

[53] Duggan, E., and S. Cherian. 2004. Integrating nominal group technique and joint application development for improved systems requirements determination. *Information & Management* 41 (4):399–411.

[54] Chiclana, F., and F. Herrera. 1998. Integrating three representation models in fuzzy multipurpose decision-making based in fuzzy preference relations. *Fuzzy Sets and Systems* 97 (1):33–48.

[55] Delgado, M., F. Herrera, and V. Herrera. 1998. Combining numerical and linguistic information in group decision-making. *Information Sciences* 107 (1):177–94.

[56] Fan, Z.-p., and Y.-p. Jiang. 2003. Approach to group decision-making with different forms of preference information based on OWG operators. *Journal of Management Sciences in China* 6 (1):32–36.

[57] Xiao, S.-h., Z.-p. Fan, and M.-g. Wang. 2000. Goal programming model for integrating two forms of preference information in group decision making. *Journal of Northeastern University* 21 (4):453–55.

[58] Chen, H.-y., and C.-l. Liu. 2005. Relative entropy aggregation method in group decision making based on different types of preference information. *Journal of Southeast University (Natural Science Edition)* 35 (2):311–15.

[59] Yager, R. Uncertainty modeling and decision support. *Reliability Engineering & System Safety* 85 (1):341–54.

[60] Xu, Z.-s. 2004. *Uncertain multiple attribute decision making methods and applications*. Beijing: Tsinghua University Press.

[61] Busemeyer, J., R. Jessup, and J. Johnson. 2006. Building bridges between neural models and complex decision making behavior. *Neural Networks* 19:1047–58.

[62] Bonaccio, S., and S. Reeshad. 2006. Advice taking and decision-making: An integrative literature review and implications for the organizational sciences. *Organizational Behavior and Human Decision Processes*. 101:127–51.

[63] Mikhailov, L. 2004. A fuzzy approach to deriving priorities from interval pairwise comparison judgments. *European Journal of Operational Research* 159:687–704.

[64] Wang, Y.-m., J.-b. Yang, and D.-l. Xu. 2005. A two-stage logarithmic goal programming methods for generating weights from interval comparison matrices. *Fuzzy Sets and Systems* 152 (1):475–98.

[65] Zhu, J.-j., S.-x. Liu, and M.-g. Wang. 2005. Integration of weights model of interval numbers comparison matrix. *Acta Automatica Sinica* 31 (3):434–39.

[66] Zhu, J., M.-g. Wang, and S.-x. Liu. 2005. Research on and application of new uncertainty model in analytical hierarchy process. *Journal of Management Sciences in China* 8 (5):15–20.

[67] Herrera, F., L. Martinez, and P. Sanchez. 2005. Managing non-homogeneous information in group decision making. *European Journal of Operational Research* 166:115–32.

[68] Wu, J. 2004. *Study on some problems in multiple attribute decision making based on interval complementary judgement matrices.* Sichur: Doctorate Dissertation of Southwest Iiao Tong University.

[69] Xu, Z.-s. 2006. Group decision making method based on different types of incomplete judgment matrices. *Control and Decision* 21 (1):28–33.

[70] Zhu, J., and S.-f. Liu. 2006. Mechanism of fuzzy preference translation in group decision. *Control and Decision* 23 (1):56–59.

[71] Zhu, J. 2006. Group aggregation approach of two kinds of uncertain preference information. *Control and Decision* 21 (8):889–92.

[72] Zhu, J., S.-f. Liu, and H.-h. Wang. 2007. Aggregation approach of two kinds of three-point interval number comparison matrix in group decision making. *Acta Automatica Sinica* 33 (3):297–301.

[73] Fan, Z., and S. Xiao. 1995. Weight method on dynamic decision making problems. *Calculation Technology and Automatization* 14 (1):20–24.

[74] Bendoly, E., and D. Bachrach. 2003. A process-based model for priority convergence in multi-period group decision-making. *European Journal of Operational Research* 148 (3):534–45.

[75] Wang, Y. 1997. Time property multiple criteria ideal point decision making method. *Soft Science of China* 7:94–98.

[76] Wang, J.-q. 1999. Gray correlation analysis of the dynamic multiple attribute decision making. *Journal of Central South University of Technology (Natural Science)* 30 (5):548–50.

[77] Xie, D.-f., W. Chen, and J. Chen. 2005. Ideal matrix method and its application to the dynamic multiple index decision making. *Science-Technology and Management* (3):40–41.

[78] Ma, J. 2005. A multi-indexes dynamic weight fuzzy model for sequential decision of water resources and its application. *Journal of Inner Mongolia Agricultural University (Natural Science Edition)* 26(1):69–74.

[79] Finan, J., and W. Hurley. 1997. Analytic hierarchy process: Does adjusting a pair wise comparison matrix to improve the consistency ratio help? *Computers & Operations Research* 24 (8):749–55.

[80] Xu, S. 1986. *Theory of the analytical hierarchy process.* Tianjin: Tianjin University Publishing House.

[81] Saaty, T. 2003. Decision making with the AHP: Why is the principal eigen-vector necessary. *European Journal of Operational Research* 145 (1):85–91.

[82] Tam, M., and V. Tummala. 2001. An application of the AHP in vendor selection of a telecommunications system. *Omega* 29 (2):171–82.

[83] Wei, Y., J. Liu, and X. Wang. 1994. Concept of consistence and weights of the judgement matrix in the uncertain type of AHP. *Systems Science and Systems Engineering* 22 (4):16–22.

[84] Bryson, N., and A. Joseph. 2000. Generating consensus priority interval vectors for group decision-making in the AHP. *Journal of Multi-Criteria Decision Analysis* (4):127–37.

[85] Leung, L., and D. Cao. 2000. On consistency and ranking of alter-natives in fuzzy AHP. *European Journal of Operational Research* 124 (1):102–13.

[86] Haines, L. 1998. A statistical approach to the analytic hierarchy process with interval judgments (I) distributions on feasible regions. *European Journal of Operational Research* 110 (1):112–25.

[87] Wang, L., G. He, and J. Li. 1997. Convex cone model for interval judgments in the analytic hierarchy process. *Journal of Systems Engineering* 12 (3):39–48.

[88] Lipovetsky, S., and A. Tishler. 1999. Interval estimation of priorities in the AHP. *European Journal of Operational Research* 114 (1):153–64.

[89] Byeong, S. 2000. The analytic hierarchy process in an uncertain envi-ronment: A simulation approach by Hauser and Tadikamalla (1996). *European Journal of Operational Research* 124 (1):217–18.

[90] Mikhailov, L. 2002. Fuzzy analytical approach to partnership selection in formation of virtual enterprises. *Omega* 30 (5):393–401.

[91] Buckley, J., T. Feuring, and Y. Hayashi. 2001. Fuzzy hierarchical analysis revisited. *European Journal of Operational Research* 129 (1):48–64.

[92] Chen, S. 1996. Evaluating weapon systems using fuzzy arithmetic opera-tions. *Fuzzy Sets and Systems* 77 (3):265–76.

[93] Kaslingam, R., and C. Lee. 1996. Selection of vendors—A mixed integer programming approach. *Computers and Industrial Engineering* 31:347–50.

[94] Degraeve, Z., E. Labro, and F. Roodhooft. 2000. An evaluation of vendor selection models from a total cost of ownership perspective. *European Journal of Operational Research* 125 (1):34–58.

[95] Talluri, S., R. Baker, and J. Sarkis. 1999. A framework for designing effi-cient value chain networks. *International Journal of Production Economics* 62 (1–2):133–44.

[96] Masood, A. A combined AHP-GP model for quality control systems. *International Journal of Production Economics* 72 (1):27–40.

[97] Ghodsypour, S., and C. Brien. 1998. A decision support system for supplier selection using an integrated analytic hierarchy process and linear pro-gramming. *International Journal of Production Economics* 56–57:199–212.

[98] Saaty, T., and L. Vargas. 1987. Uncertainty and rank order in the AHP. *European Journal of Operational Research* 32:107–17.

[99] Xu, Z.-s. 2001. Algorithm for priority of fuzzy complementary judge-ment matrix. *Journal of Systems Engineering* 16 (4):311–14.

[100] Dong, Y., Y. Xu., and H. Li. 2008. On consistency measures of linguistic preference relations. *European Journal of Operational Research* 189:430–44.

[101] Dong, Y. C., Y. F. Xu, and S. Yu. 2009. Linguistic multi-person decision making based on the use of multiple preference relations. *Fuzzy Sets and Systems* 160:603–23.

[102] Vahdani, B., A. H. K. Jabbari, V. Roshanaei, and M. Zandieh. 2010. Extension of the ELECTRE method for decision-making problems with interval weights and data. *International Journal of Advanced Manufacturing Technology* 50 (5–8):793–800.

[103] Almeida-Dias, J., J. R. Figueira, and B. Roy. 2010. ELECTRE TRI-C: A multiple criteria sorting method based on characteristic reference actions. *European Journal of Operational Research* 204 (3):565–80.

[104] Chen, C. T., P. F. Pai, and W. Z. Hung. 2010. An integrated methodology using linguistic PROMETHEE and maximum deviation method for third-party logistics supplier selection. *International Journal of Computational Intelligence Systems* 3 (suppl. SI, 4):438–51.

[105] Li, W. X. and B. Y. Li. 2010. An extension of the Promethee II method based on generalized fuzzy numbers. *Expert Systems with Applications* 37 (7):5314–19.

[106] Liu, S. F., N. M. Xie, and J. Forrest. 2010. On new models of grey incidence analysis based on visual angle of similarity and nearness. *Systems Engineering and Theory & Practice* 30:881–87.

[107] Liu, S. F., and Y. Lin. 2006. *Grey information: Theory and practical applications*. London: Springer-Verlag.

[108] Fan, Z. P., and Y. Liu. 2010. A method for group decision-making based on multi-granularity uncertain linguistic information. *Expert Systems with Applications* 37:4000–4008.

[109] Xu, Y. J., and Q. L. Da. 2009. Approach based on multi-granularity linguistic judgment matrices in multi-attribute group decision making. *Journal of Industrial Engineering Management* 23:69–73.

[110] Cabrerizo, F. J., I. J. Perez, and E. Herrera-Viedma. Managing the consensus in group decision making in an unbalanced fuzzy linguistic context with incomplete information. *Knowledge-Based Systems* 23:169–81.

[111] Wang, T. C., and Y. H. Chen. 2008. Applying fuzzy linguistic preference relations to the improvement of consistency of fuzzy AHP. *Information Sciences* 178:3755–65.

[112] Lahdelma, R., and Salminen Pekka. 2009. Prospect theory and stochastic multicriteria acceptability analysis (SMAA). *Omega* 37:961–71.

[113] Malpica, J. A., M. C. Alonso., and M. A. Sanz. 2007. Dempster–Shafer theory in geographic information systems: A survey. *Expert Systems with Applications* 32 (1):47–55.

[114] Basir, O., and X. Yuan. 2007. Engine fault diagnosis based on multi-sensor information fusion using Dempster–Shafer evidence theory. *Information Fusion* 8 (4):379–86.

[115] Wu, D. 2009. Supplier selection in a fuzzy group setting: A method using grey related analysis and Dempster–Shafer theory. *Expert Systems with Applications* 36 (5):8892–99.

[116] Domínguez-García, A. D., and J. G. Kassakian. 2008. An integrated methodology for the dynamic performance and reliability evaluation of fault-tolerant systems. *Reliability Engineering & System Safety* 93:1628–49.

[117] Zhao, Y. J., and G. Feng. 2009. Dynamic formulation and performance evaluation of the redundant parallel manipulator. *Robotics and Computer-Integrated Manufacturing* 25:770–81.

[118] Guo, Y. J. 2007. *Integration evaluation theory method and application.* Beijing: Science Press.

[119] Sun, C.-C. 2010. A performance evaluation model by integrating fuzzy AHP and fuzzy TOPSIS methods. *Expert Systems with Applications* 37:7745–54.

[120] Ye, F. 2010. An extended TOPSIS method with interval-valued intuitionistic fuzzy numbers for virtual enterprise partner selection. *Expert Systems with Applications* 37:7050–55.

[121] Chu, T.-C., and Y.-C. Lin. 2009. Interval arithmetic based fuzzy TOPSIS model. *Expert Systems with Applications* 36:10870–76.

[122] Sadi-Nezhad, S., and K. K. Damghani. 2010. Application of a fuzzy TOPSIS method base on modified preference ratio and fuzzy distance measurement in assessment of traffic police centers performance. *Applied Soft Computing* 10:1028–39.

[123] Roghanian, E., J. Rahimi, and A. Ansari. 2010. Comparison of first aggregation and last aggregation in fuzzy group TOPSIS. *Applied Mathematical Modelling* 34:3754–66.

[124] Jahanshahloo, G. R., F. Hosseinzadeh Lotfi, and A. R. Davoodi. 2009. Extension of TOPSIS for decision-making problems with interval data: Interval efficiency. *Mathematical and Computer Modelling* 49:1137–42.

[125] Wang, Y.-J., and H.-S. Lee. 2007. Generalizing TOPSIS for fuzzy multiple-criteria group decision-making. *Computers and Mathematics with Applications* 53:1762–72.

[126] Shih, H.-S. 2008. Incremental analysis for MCDM with an application to group TOPSIS. *European Journal of Operational Research* 186:720–34.

[127] Mohammad, I. 2009. Using the Hamming distance to extend TOPSIS in a fuzzy environment. *Journal of Computational and Applied Mathematics* 231:200–207.

[128] Chou, S. Y., Y. H. Chang, and C. Y. Shen. 2008. A fuzzy simple additive weighting system under group decision-making for facility location selection with objective/subjective attributes. *European Journal of Operational Research* 189:132–45.

[129] Xu, Z. 2005. Deviation measures of linguistic preference relations in group decision making. *Omega* 33:249–54.

[130] Song, G.-x., and D.-l. Yang. 2003. An improved consistency method of checking the fuzzy comparison matrix. *Systems Engineering* 21 (1):110–16.

[131] Apostolou, B. 2002. Note on consistency ratio: A reply. *Mathematical and computer modeling* 35 (9–10):1081–1083.

Index

A

Aggregation method of hybrid
uncertain comparison
matrix. *See also* Hybrid
uncertain comparison index
background, 125–126, 135–136
case example, 146
definitions, 136
group decision making, general
idea of, for, 140–142
multiplicative consistency, 127
overview, 126–127, 136–137
two kinds of preference
information, considering
consistency coordination,
160–161
weight of, 127
Aggregation method of reciprocal and
complementary comparison
matrix, multiple-stage
alternatives, ranking, 178
decision maker, weight modeling,
178–179
deviations of group and
individual preferences, 181

overview, 177–178
process of, 187–189
reciprocal judgment matrix,
establishing, 178
stages, weights of, 188–189
Aggregation method of uncertain
preferences with multiple
stages, 191
case example, 194–196,
197–199
decision maker, weight of,
189–190
information quality, 199
time latitude, 199
uncertain preference, weight of,
189
weight model, based on least
deviation, 192–193
Analytical hierarchy process (AHP)
fuzzy number, 50
information loss, 51–52
interval number, 50
limits of, in group decisions,
51, 67
random, 50
random crisp matrix, 18

unascertained number
comparison matrix; *See*
Unascertained number
comparison matrix
use in multiple criteria decision
making, 13, 47
vendor selection process,
combined with, 26–27
weighting, 63
Arrow, K. J., 4, 6

B

Bayesian theory, 3
Blind number AHP comparison
matrix, 55

C

Cellular automata, 6
Complementary judgment matrix,
36, 40, 41, 42, 45, 46, 47,
162, 163, 165
Consistency ratio, 162
Convex programming, 181
Cybernetics, 6

D

Decision makers
acceptable lack of satisfaction
boundaries, 129–130
adjustments of mathematical
meaning according to
preferences, 101
aggregation of multivariate
uncertain decision makers, 10
inconsistencies, 133
objective judgment, 47–48
preferences of, 7, 181, 213
subjective preference, 47–48
weights of, 167–168, 178–179,
182, 187, 213, 222

Decision making
accuracy, 130
asymmetry of, 2
complexity of, 2, 56, 88, 93, 128
defining, 1
group; *See* Group decision
making
multistage, 3
origins of research on, 3
stages of, 2
subjectivity, role of, 1
uncertainty, role of, 1, 2–3, 4;
See also Uncertain decision
making
Decision theory, 3
Determinate comparison matrix, 55
Deviation of decisions, 179

E

Earned value management
performance evaluation
system, 169–170, 171–172,
173, 175
Eigenvector method, 22
ELECTRE, 201
Enterprise resource planning
(ERP), 26

F

Fisher, R. A., 3
Fuzzy analytical hierarchy process
(AHP), 66
Fuzzy group decision theory, 5
Fuzzy inequality, 128
Fuzzy math, 3
Fuzzy numbers
comparison matrix for, 14
defuzzifying, 203
operation rule of classifying, 210
Fuzzy set-based decision making,
3–4

G

Geometry average method, 51
Gray clustering, 10
  case example, 158–160
  consistency, achieving, 155
  coordination of decision makers'
    adjustments, 156
  defining, 149–150
  fuzzy preferences, with, 150
  large-scale group decisions based
    on, 152–153
  meta syntheses of decision
    group preference between
    classification, 156–157
  overview, 148–149
  reclassification of group
    preferences, 149
  unified weight approach of
    multiple uncertainty
    preferences, 150–152
Gray number, 49
Gray set, 49
Gray systems, 3
Group decision making
  aggregation model for, 140–142
  communications between
    members, 5
  complexity of, 6–7, 8, 88
  consistency modification model
    for, 161, 162
  democratic processes, 175
  dynamic process of, 5
  organization theory of; *See*
    Organization theory
    of group decision
    making
  preference aggregation, 5, 7–9
  research on, 3–4
  strategic behavior in, 5
  support system platform, 5
  synthesis preferences, 143,
    144–145

H

Hybrid uncertain comparison index,
  174
  case example, 134–135
  consistency, 127, 128, 131, 132
  feasible regions, 132–133
  overview, 126–127
  weight of, 127
  weight solution method, 127, 128

I

Impossibility theorem, 6
Information distortion, 8
Interval number comparison matrix
  (INCM), 62, 227
  consistency, 14
  consistency analysis, 13, 20,
    47–48
  genetic algorithm, 24–25
  integration model, 23–24
  local consistency, 14, 15
  local satisfactory consistency, 15,
    16, 18, 21
  random crisp comparison matrix,
    18, 19
  tolerance deviations, 15, 16
  total consistency, 17
  weight distribution, 39, 44
  weight model, 20
  weight range model, 20, 21
  weights feasible region, 14, 17
Interval number complementary
    judgment matrix, 44
Interval number judgment matrix, 44
Interval number preference
    sequence, 228

J

Jarque-Bera method, 94
Judgment matrices, 173

L

Linguistic comparison matrix
  accuracy, 98
  aggregation method, on multiple-
    criteria decision, with
    duality semantic transform,
    110–111, 112, 113,
    114–115, 116, 117
  case example, 119–120, 121–122
  comprehensive index value, 118
  consistency analysis, 93, 94
  decision maker preferences, 117
  decision maker weights, 118, 119
  defining, 92, 93
  group aggregation property, 102
  language phrase sets, 92
  modification method, 98–99
  multiple comparisons, 96
  order consistency test approach,
    103, 109
  overview, 91–92
  problems inherent in, 123–124
  satisfaction consistency index, 96,
    97–98, 101, 105, 106–108,
    109, 110
  transitive linguistic preference
    and adjustment, 103, 110
Logistics management, 25–26

M

Multiplicative consistency, 127

N

National Science Foundation, U.S., 6
Neural networks, 6

O

Organization theory of group
  decision making, 5

P

Probability theory, 3
PROMETHEE, 201
Purchasing logistics, 25

R

Raiffa, H., 3
Random analytical hierarchy process
  (AHP), 66
Random deterministic judgment
  matrix, 37, 39
Random group decision theory, 5
Rank correlation coefficient,
  Spearman's, 191
Rank interval number based on
  possible method, 45
Reachability matrix, 103
Reciprocal judgment matrix, 163,
  166, 178, 228
Reciprocal matrix, 22
Rough sets, 3
Row sum normalization, 36, 42

S

Santa Fe Institute, 6
Savage, J. L., 3
Schlaifer, R. O., 3
Side lengths, 103–104
Sorting method of interval numbers
  weight, 134
Spearman's rank correlation
  coefficient, 191

T

Three-point interval number
  comparison matrix
  aggregation of two kind, 73–74,
    75, 76
  application of in process decision
    making, 67–68

base paper specification reduction
  example, 79–85, 85–88
consistency analysis, 69, 70,
  71–72, 76
defining, 66–67
deviation distances, 70–71, 72
economic analysis, 85
judgment interval, 68
management analysis, 85
multiple comparison methods,
  80–81
overview, 65–66
primary indexes, 85
quality analysis, 85
secondary indexes, 85–86
weight analysis, 78–79, 85
weighting estimates, 69
TOPSIS, 201
  alternatives, ranking, 205–207,
    208–209, 213
  applications, 202
  case example, 215, 217
  decision makers, weighting, 213,
    222
  fuzzy numbers, 203, 204
  incomplete weight of DM and
    stage with multistage
    lingustic evaluation for
    group decision, 210–213
  linguistic sets, 203
  linguistic variables, measuring
    distance of, 202
  negative ideal point, 204, 205,
    206, 209, 210, 217
  positive ideal point, 204, 205,
    206, 209, 210, 217
  practicality, 202
  reliability, 202
  stages, weighting, 222
  weight for criteria and decisions
    makers and stages with
    incomplete information,
    218–219, 222

U

Unascertained mathematics, 3
Unascertained number analytical
  hierarchy process
  (UNAHP), 55
  Monte Carlo weight model, 59–60
  weight calculations, 59
Unascertained number comparison
  matrix
  binary arrays, use of, 52, 53
  consistency analysis, 56–57
  defining, 53–54
  discovery of, 49
  discrete distribution, 53, 64
  expansion of AHP, 50–51
  group decision making, use in, 64
  high-risk unascertained numbers,
    58–59
  judging credibility, 53
  operation rule of unascertained
    number, 57–59
  satisfaction analysis, 56–57
  weight integration based on
    Monte Carlo, 60, 61, 62
  weight integration based on rule
    of unascertained number,
    60, 62
Unascertained rational number
  comparison matrix, 54
Uncertain decision making
  aggregation of multivariate
    uncertain decision makers,
    10
  consistent conversion
    mechanisms, 8–9
  internal decision-making
    mechanisms, 8
  research on, 3
  uncertain preference information, 9
  uncertain preference information
    with timing characteristics,
    11

Uncertain judgment matrix, 227
Uncertain ordered weighed
   averaging, 228
Uniform distribution probability,
   37
Utility value, 227

V

Vendor selection approaches, 25–27
   comparison matrices,
      31–32
   comprehensive weights, 33,
      34
   cost performance criteria, 28, 34
   criteria weights, 29
   delivery ability, 29
   quality criteria, 28–29, 34
   service level, 29
   subcriteria weights, 29–31

W

Wald, Abraham, 3
Weight estimation analysis, 13–14
Weight solution models, 48
Weight-adding method, 51
Weighted arithmetic average
   method, 164
Weights modeling
   interval number complementary
      judgment matrices, for,
      138, 139
   interval number preference
      sequence, for, 139–140
   interval number reciprocal
      judgment, for, 137–138
   value of utility of interval
      number, for, 139
World Trade Organization (WTO),
   25